BRILLIANT BLUNDERS

From Darwin to Einstein—
Colossal Mistakes by Great Scientists
That Changed Our Understanding of
Life and the Universe

Mario Livio

Simon & Schuster

NEW YORK · LONDON · TORONTO · SYDNEY · NEW DELHI

Simon & Schuster
1230 Avenue of the Americas
New York, NY 10020

First Simon & Schuster hardcover edition May 2013

SIMON & SCHUSTER and colophon are registered trademarks
of Simon & Schuster, Inc.

Credits for figures and quotations are listed on pp. 325–26.

For information about special discounts for bulk purchases,
please contact Simon & Schuster Special Sales at
1-866-506-1949 or business@simonandschuster.com.

The Simon & Schuster Speakers Bureau can bring authors
to your live event. For more information or to book an event,
contact the Simon & Schuster Speakers Bureau at
1-866-248-3049 or visit our website at www.simonspeakers.com.

Text designed by Paul Dippolito

Manufactured in the United States of America

7   9   10   8   6

Library of Congress Cataloging-in-Publication Data

Livio, Mario.
Brilliant blunders : from Darwin to Einstein — colossal mistakes by great
scientists that changed our understanding of life and the universe / Mario Livio.
— First Simon & Schuster hardcover edition.
    p. cm.
Summary: "Drawing on the lives of five great scientists — Charles Darwin,
William Thomson (Lord Kelvin), Linus Pauling, Fred Hoyle and Albert
Einstein — scientist/author Mario Livio shows how even the greatest scientists
made major mistakes and how science built on these errors to achieve
breakthroughs, especially into the evolution of life and the universe." —
Provided by publisher.
Includes bibliographical references and index.
1. Errors, Scientific. I. Title.
    Q172.5.E77L58 2013
    500 — dc23   2012037732

ISBN 978-1-4391-9236-8
ISBN 978-1-4391-9238-2 (ebook)

To Noga and Danielle

CONTENTS

Preface   *1*

1. Mistakes and Blunders   *5*

2. The Origin   *12*

3. Yea, All Which It Inherit, Shall Dissolve   *37*

4. How Old Is the Earth?   *60*

5. Certainty Generally Is Illusion   *84*

6. Interpreter of Life   *103*

7. Whose DNA Is It Anyway?   *136*

8. *B* for Big Bang   *157*

9. The Same Throughout Eternity?   *184*

10. The "Biggest Blunder"   *221*

11. Out of Empty Space   *246*

Coda   *269*

Notes   *273*

Bibliography   *303*

Index   *327*

PREFACE

===

Throughout the entire period that I have been working on this book, every few weeks someone would ask me what my book was about. I developed a standard answer: "It is about blunders, and it is *not* an autobiography!" This would get a few laughs and the occasional approbation "What an interesting idea." My objective was simple: to correct the impression that scientific breakthroughs are purely success stories. In fact, nothing could be further from the truth. Not only is the road to triumph paved with blunders, but the bigger the prize, the bigger the potential blunder.

Immanuel Kant, the great German philosopher, wrote famously, "Two things fill the mind with ever new and increasing admiration and awe, the more often and steadily we reflect upon them: the *starry heavens above me and the moral law within me.*" In the time that has passed since the publication of his *The Critique of Practical Reason* (1788), we have made impressive progress in understanding the former; considerably less so, in my humble opinion, in elucidating the latter. It is apparently much more difficult to make life or mind comprehensible to itself. Nevertheless, the life sciences in general—and the research into the operation of the human brain in particular—are truly picking up speed. So it may not be altogether inconceivable after all that one day we will even fully understand why evolution has concocted a sentient species.

While this book is about some of the remarkable endeavors to figure out life and the cosmos, it is more concerned with the journey than with the destination. I tried to concentrate on the thought process and the obstacles on the way to discovery rather than on the achievements themselves.

Many people have helped me along the way, some maybe even unknowingly. I am grateful to Steve Mojzsis and Reika Yokochi for discussions on topics related to geology. I thank Jack Dunitz, Horace Freeland Judson, Matt Meselson, Evangelos Moudrianakis, Alex Rich, Jack Szostak, and Jim Watson for conversations on chemistry, biology, and specifically on Linus Pauling's work. I am indebted to Peter Eggleton, John Faulkner, Geoffrey Hoyle, Jayant Narlikar, and Lord Martin Rees for helpful discussions on astrophysics and cosmology, and on Fred Hoyle's work.

I would also like to express my gratitude to all the people who provided me with invaluable materials for this book, and in particular to: Adam Perkins and the staff of the Cambridge University Library, for materials on Darwin and on Lord Kelvin; Mark Hurn of the Institute of Astronomy, Cambridge, for materials on Lord Kelvin and on Fred Hoyle; Amanda Smith of the Institute of Astronomy, Cambridge, for materials on Fred Hoyle and for processing photos related to Watson and Crick; Clifford Meade and Chris Petersen of the Special Collections Department of Oregon State University, for materials on Linus Pauling; Loma Karklins of the Caltech Archives, for material on Linus Pauling; Sarah Brooks from the Nature Publishing Group, for material on Rosalind Franklin; Bob Carswell and Peter Hingley for materials on Georges Lemaître from the Royal Astronomical Society; Liliane Moens of the Archives Georges Lemaître, for materials on Georges Lemaître; Kathryn McKee of St. John's College, Cambridge, for materials on Fred Hoyle; and Barbara Wolff of the Albert Einstein Archives, Diana Kormos Buchwald of the Einstein Papers Project, Daniel Kennefick of the University of Arkansas, Michael Simonson of the Leo Baeck Institute, Christine Lutz of Princeton University, and Christine Di Bella of the Institute for Advanced Study for materials on Einstein.

Special thanks are due to Jill Lagerstrom, Elizabeth Fraser, and Amy Gonigam of the Space Telescope Science Institute, and to the staff at the Johns Hopkins University Library for their continuous bibliographic support. I am grateful to Sharon Toolan for her professional help in preparing the manuscript for print, to Pam Jeffries for skillfully drawing some of the figures, and to Zak Concannon for cleaning some of the figures. As always, my most patient and supportive ally has been my wife, Sofie.

Finally, I thank my agent, Susan Rabiner, for her relentless encouragement; my editor, Bob Bender, for his thoughtful comments; Loretta Denner, for her assistance during copyediting; and Johanna Li, for her dedication during the entire production of this book.

CHAPTER 1

MISTAKES AND BLUNDERS

Great blunders are often made, like large ropes, of a multitude of fibres. Take the cable thread by thread, take separately all the little determining motives, you break them one after another, and you say: that is all. Wind them and twist them together they become an enormity.

—VICTOR HUGO, *LES MISÉRABLES*

When the mercurial Bobby Fischer, perhaps the most famous chess player in the history of the game, finally showed up in Reykjavik, Iceland, in the summer of 1972 for his world championship match against Boris Spassky, the anticipation in the chess world was so thick you could cut it with a chain saw. Even people who had never shown any interest in chess before were holding their breath for what had been dubbed "the Match of the Century." Yet in the twenty-ninth move of the very first game, in a position that appeared to be leading to a dead draw, Fischer chose a move that even amateur chess players would have rejected instinctively as a mistake. This may have been a typical manifestation of what is known as "chess blindness"—an error that in the chess literature is denoted by "??"—and would have disgraced a five-year-old in a local chess club. Particularly astonishing was the fact that the mistake was committed by a man who'd smashed his way to the match with the Russian Spassky after an extraordinary sequence of twenty successive wins against the world's top players. (In most world-class competitions, there are easily as many draws as outright victories.) Is this type of "blindness" something that happens only in chess? Or are other intellectual enterprises also prone to similarly surprising mistakes?

Oscar Wilde once wrote, "Experience is the name everyone gives to their mistakes." Indeed, we all make numerous mistakes in our everyday lives. We lock our keys inside the car, we invest in the wrong stock (or sometimes in the right stock, but at the wrong time), we grossly overestimate our ability to multitask, and we often blame the absolutely wrong causes for our misfortunes. This misattribution, by the way, is one of the reasons that we rarely actually learn from our mistakes. In all cases, of course, we realize that these were mistakes only after we have made them—hence, Wilde's definition of "experience." Moreover, we are much better at judging other people than at analyzing ourselves. As psychologist and Nobel laureate in economics Daniel Kahneman has put it, "I am not very optimistic about people's ability to change the way they think, but I am fairly optimistic about their ability to detect the mistakes of others."

Even attentively and carefully constructed processes, such as those involved in the criminal justice system, fail occasionally—sometimes heartbreakingly so. Ray Krone of Phoenix, Arizona, for instance, spent more than ten years behind bars and faced the death penalty after having been convicted *twice* of a brutal murder he did not commit. He was eventually fully exonerated (and the real killer implicated) by DNA evidence.

The focus of this book, however, is not on such mistakes, no matter how grave they may be: it is on major *scientific blunders*. By "scientific blunders," I mean particularly serious conceptual errors that could potentially jeopardize entire theories and game plans, or could, in principle at least, hold back the progress of science.

Human history teems with stories of momentous blunders in a wide range of disciplines. Some of these consequential errors go all the way back to the Scriptures, or to Greek mythology. In the book of Genesis, for instance, the very first act of Eve—the biblical mother of all living humans—was to yield to the crafty serpent and to eat the forbidden fruit. This monumental lapse in judgment led to no less than the banishment of Adam and Eve from the Garden of Eden, and—at least according to the thirteenth-century theologian Thomas Aquinas—even to humans being eternally denied access to absolute truth. In the Greek mythology, Paris's misguided elopement with the beautiful Helen, the wife of the king of Sparta, brought about the total destruction of the city of Troy. But these examples don't even begin to scratch the surface. Throughout history, neither renowned military commanders nor famous philosophers or groundbreaking thinkers were immune to serious blunders. During World War II, the German field marshal Fedor von Bock foolishly repeated Napoléon's ill-fated attack on Russia in 1812. Both officers failed to appreciate the insurmountable powers of "General Winter"—the long and harsh Russian winter for which they were woefully unprepared. The British historian A. J. P. Taylor once summarized Napoléon's calamities this way: "Like most of those who study history, he [Napoléon] learned from the mistakes of the past how to make new ones."

In the philosophical arena, the great Aristotle's erroneous ideas on physics (such as his belief that all bodies move toward their "natural" place) fell just as wide off the mark as did Karl Marx's awry predictions on the imminent collapse of capitalism. Similarly, many of Sigmund Freud's psychoanalytic speculations, be it on the "death instinct"—a supposed impulse to return to a pre-life state of quietude—or on the role of an infantile Oedipus complex in the neuroses of women, have been found to be pathetically amiss, to put it mildly.

You may think, OK, people made mistakes, but surely, when it comes to some of the greatest *scientists* of the past two centuries—such as the twice Nobel laureate Linus Pauling or the formidable Albert Einstein—they were correct at least in those theories for which they are best known, right? After all, hasn't the intellectual glory of modern times been precisely in the establishment of science as an empirical discipline, and of error-proof mathematics as the "language" of fundamental science? Were, then, the theories of these illustrious minds and of other comparable thinkers truly free of serious blunders? Absolutely not!

The purpose of this book is to present in detail some of the surprising blunders of a few genuinely towering scientists, and to follow the unexpected consequences of those blunders. At the same time, my goal is also to attempt to analyze the possible causes for these blunders and, to the extent possible, to uncover the fascinating relations between those blunders and features or limitations of the human mind. Ultimately, however, I hope to demonstrate that the road to discovery and innovation can be constructed even through the unlikely path of blunders.

As we shall see, the delicate threads of evolution interweave all the particular blunders that I have selected to explore in detail in this book. That is, these are serious blunders related to the theories of the evolution of life on Earth, the evolution of the Earth itself, and the evolution of our universe as a whole.

Blunders of Evolution and Evolution of Blunders

One of the definitions of the word "evolution" in the *Oxford English Dictionary* reads: "The development or growth, according to its inherent tendencies, of anything that may be compared to a living organism . . . Also, the rise or origination of anything by natural development, as distinguished from its production by a specific act." This was not the original meaning of the word. In Latin, *evolutio* referred to the unrolling and reading of a book that existed in the form of a scroll. Even when the word started to gain popularity in biology, it was used initially only to describe the growth of an embryo. The first utilization of the word "evolution" in the context of the genesis of species can be found in the writings of the eighteenth-century Swiss naturalist Charles Bonnet, who argued that God had pre-organized the birth of new species in the germs of the very first life-forms he created.

In the course of the twentieth century, the word "evolution" has become so intimately associated with Darwin's name that you may find it hard to believe that in the first, 1859 edition of his masterwork, *On the Origin of Species*, Darwin does not mention the word "evolution" as such even once! Still, the very last word of *The Origin* is "evolved."

In the time that has passed since the publication of *The Origin*, evolution has assumed the broader meaning of the definition above, and today we may speak of the evolution of such diverse things as the English language, fashion, music, and opinions, as well as of sociocultural evolution, software evolution, and so on. (Check out how many web pages are devoted just to "the evolution of the hipster.") President Woodrow Wilson emphasized once that the correct way to understand the Constitution of the United States was through evolution: "Government is not a machine, but a living thing . . . It is accountable to Darwin, not to Newton."

My focus on the evolution of life, of the Earth, and of the universe should not be taken to mean that these are the only scientific arenas in which blunders have been committed. Rather, I have chosen these particular topics for two main reasons. First, I wanted to critically review the blunders made by some of the scholars that appear on almost everybody's short list of great minds. The blunders of such luminaries, even if of a past century, are extremely relevant to questions scientists (and, indeed, people in general) face today. As I hope to show, the analysis of these blunders forms a living body of knowledge that is not only captivating in its own right but also can be used to guide actions in domains ranging from scientific practices to ethical behavior. The second reason is simple: The topics of the evolution of life, of the Earth, and of the universe have intrigued humans—not just scientists—since the dawn of civilization, and have inspired tireless quests to uncover our origins and our past. The human intellectual curiosity about these subjects has been at least partially at the root of religious beliefs, of the mythical stories of creation, and of philosophical inquiries. At the same time, the more empirical, evidence-based side of this curiosity has ultimately given birth to science. The progress that humankind has made toward deciphering some of the complex processes involved in the evolution of life, the Earth, and the cosmos is nothing short of miraculous. Hard to believe, but we think that we can trace cosmic evolution back to when our universe was only a fraction of a second old. Even so, many questions remain unanswered, and the topic of evolution continues to be a hot-button issue even today.

It took me quite a while to decide which major scientists to include in this journey through deep intellectual and practical waters, but I eventually converged on the blunders of five individuals. My list of surprising "blunderers" includes the celebrated natural-ist Charles Darwin; the physicist Lord Kelvin (after whom a temperature scale is named); Linus Pauling, one of the most influential chemists in history; the famous English astrophysicist and cosmologist Fred Hoyle; and Albert Einstein, who needs no introduction. In each case, I will address the central theme from two rather different—but complementary—perspectives. On one hand, this will be a book about some of the theories of these great savants and the fascinating relations among those theories, viewed in part from the unusual vantage point of their weaknesses and sometimes even failures. On the other, I will scrutinize briefly the various types of blunders and attempt to identify their psychological (or, if possible, neuroscientific) causes. As we shall see, blunders are not born equal, and the blunders of the five scientists on my list are rather different in nature. Darwin's blunder was in not realizing the full implications of a particular hypothesis. Kelvin blundered by ignoring unforeseen possibilities. Pauling's blunder was the result of overconfidence bred by previous success. Hoyle erred in his obstinate advocacy of dissent from mainstream science. Einstein failed because of a misguided sense of what constitutes aesthetic simplicity. The main point, however, is that along the way, we shall discover that blunders are not only inevitable but also an essential part of progress in science. The development of science is not a direct march to the truth. If not for false starts and blind alleys, scientists would be traveling for too long down too many wrong paths. The blunders described in this book have all, in one way or another, acted as catalysts for impressive breakthroughs—hence, their description as "brilliant blunders." They served as the agents that lifted the fog through which science was progressing, in its usual succession of small steps occasionally punctuated by quantum leaps.

I have organized the book in such a way that for each scientist, I first present the *essence* of some of the theories for which this individual is best known. These are very concise summaries intended to provide an introduction to the ideas of these masters and an appropriate context for the blunders, rather than to represent comprehensive descriptions of the respective theories. I have also chosen to concentrate only on *one* major blunder in each case instead of reviewing a laundry list of every possible mistake that these pundits may have committed during their long careers. I shall start with the man about whom the *New York Times* correctly wrote in its obituary notice (published on April 21, 1882) that he "has been read much, but talked about more."

CHAPTER 2

THE ORIGIN

There is grandeur in this view of life, with its several powers, having been originally breathed into a few forms or
into one; and that, whilst this planet has gone cycling on
according to the fixed law of gravity, from so simple a
beginning endless forms most beautiful and most wonderful have been, and are being, evolved.

— CHARLES DARWIN

The most striking thing about life on Earth is its prodigious
diversity. Take a casual stroll on a spring afternoon; you
are likely to encounter several kinds of birds, many insects,
perhaps a squirrel, a few people (some may be walking their dogs),
and a large variety of plants. Even just in terms of the properties
that are the easiest to discern, organisms on Earth differ in size,
color, shape, habitat, food, and capabilities. On one hand, there are
bacteria that are less than one hundred thousandth of an inch in
length, and on the other, there are blue whales more than 100 feet
long. Among the thousands of known species of the marine mol-
lusks known as nudibranchs, there are many that are plain looking,
while others have some of the most sumptuous colors exhibited by
any creature on Earth. Birds can fly at astonishing heights in the
atmosphere: On November 29, 1975, a large vulture was sucked
into a jet engine at a height of 37,900 feet above the Ivory Coast in
West Africa. Other birds, such as the migrating bar-headed geese
and the whooper swans, regularly fly higher than 25,000 feet. Not
to be outdone, ocean creatures achieve similar records in depth. On

January 23, 1960, the record-setting explorer Jacques Piccard and Lieutenant Don Walsh of the US Navy descended slowly in a special probe called a bathyscaphe to the deepest point at the bottom of the Pacific Ocean—the Mariana Trench—south of Guam. When they finally touched down at the record depth of about 35,800 feet, they were amazed to discover around them a new type of bottom-dwelling shrimp that did not seem to be bothered by the ambient pressure of some 17,000 pounds per square inch. On March 26, 2012, film director James Cameron reached the deepest point in the Mariana Trench in a specially designed submersible. He described it as a gelatinous landscape as desolate as the Moon. But he also reported seeing tiny shrimp-like critters no bigger than an inch in length.

Nobody knows for sure how many species are currently living on Earth. A recent catalogue, published in September 2009, formally describes and gives official names to about 1.9 million species. However, since most living species are microorganisms or very tiny invertebrates, many of which are very difficult to access, most estimates of the total number of species are little more than educated guesses. Generally, estimates range from 5 million to about 100 million different species, although a figure of 5 to 10 million is considered probable. (The most recent study predicts about 8.7 million.) This large uncertainty is not at all surprising once we realize that just one tablespoon of dirt beneath our feet could harbor many thousands of bacterial species.

The second amazing thing characterizing life on Earth, besides its diversity, is the incredible degree of *adaptation* that both plants and animals exhibit. From the anteater's tubelike snout, or the chameleon's long and fast-moving tongue (capable of hitting its prey in about 30 thousandths of a second!), to the woodpecker's powerful, characteristically shaped beak, and the lens of the eye of a fish, living organisms appear to be perfectly fashioned for the requirements that life imposes on them. Not only are bees constructed so that they can comfortably fit into the flowering plants from which they extract nectar, but the plants themselves exploit the visits of these bees for their own propagation by polluting the bees' bodies and legs with pollen, which is then transported to other flowers.

There are many different biological species that live in an astonishing "scratch my back and I will scratch yours" interaction, or *symbiosis*. The ocellaris clown fish, for instance, dwells among the stinging tentacles of the Ritteri sea anemone. The tentacles protect the clown fish from its predators, and the fish returns the favor by shielding the anemone from other fish that feed on anemones. The special mucus on the clown fish's body safeguards it from the poisonous tentacles of its host, further perfecting this harmonious adaptation. Partnerships have even developed between bacteria and animals. For example, at seafloor hydrothermal vents, mussels bathed in hydrogen-rich fluids were found to thrive by both supporting and harvesting an internal population of hydrogen-consuming bacteria. Similarly, a bacterium from the genus *Rickettsia* was found to ensure survival advantages for the sweet potato whiteflies—and thereby for itself.

Parenthetically, one quite popular example of an astonishing symbiotic relationship is probably no more than a myth. Many texts describe the reciprocation between the Nile crocodile and a small bird known as the Egyptian plover. According to Greek philosopher Aristotle, when the crocodile yawns, the little bird "flies into its mouth and cleans his teeth"—with the plover thereby getting its food—while the crocodile "gets ease and comfort." A similar description appears also in the influential *Natural History* by the first-century natural philosopher Pliny the Elder. However, there are absolutely no accounts of this symbiosis in the modern scientific literature, nor is there any photographic record that documents such a behavior. Maybe we shouldn't be too surprised, given the rather questionable record of Pliny the Elder: Many of his scientific claims turned out to be false!

The prolific diversity, coupled with the intricate fitting together and adaptation of a wondrous wealth of life-forms, convinced many natural theologians, from Thomas Aquinas in the thirteenth century to William Paley in the eighteenth, that life on Earth required the crafting hand of a supreme architect. Such ideas appeared even as early as the first century BCE. The famous Roman orator Marcus Tullius Cicero argued that the natural world had to stem from some divine "reason":

> If all the parts of the universe have been so appointed that they could neither be better adapted for use nor be made more beautiful in appearance . . . If, then, nature's attainments transcend those achieved by human design, and if human skill achieves nothing without the application of reason, we must grant that nature too is not devoid of reason.

Cicero was also the first to invoke the clock-maker metaphor that later became the touchstone argument in favor of an "intelligent designer." In Cicero's words:

> It can surely not be right to acknowledge as a work of art a statue or a painted picture, or to be convinced from distant observations of a ship's course that its progress is controlled by reason and human skills or upon examination of the design of a sundial or a water-clock to appreciate that calculation of the time of day is made by skill and not by chance, yet none the less to consider that the universe is devoid of purpose and reason, though it embraces those very skills, and the craftsmen who wield them, and all else beside.

This was precisely the line of reasoning adopted by William Paley almost two millennia later: A contrivance implies a contriver, just as a design implies a designer. An intricate watch, Paley contended, attests to the existence of a watchmaker. Therefore, shouldn't we conclude the same about something as exquisite as life? After all, "Every indication of contrivance, every manifestation of design, which existed in the watch, exists in the works of nature; with the difference, on the side of nature, of being greater and more, and that in a degree which exceeds all computation." This fervent pleading for the imperative need for a "designer" (since the only possible but unacceptable alternative was considered to be fortuitousness or chance) convinced many natural philosophers until roughly the beginning of the nineteenth century.

Implicit in the design argument was yet another dogma: Species were believed to be absolutely *immutable.* The idea of eternal existence had its roots in a long chain of convictions about other entities that were considered enduring and unchanging. In the Aristotelian tradition, for instance, the sphere of the fixed stars was assumed to be totally inviolable. Only in Galileo's time was this particular notion completely shattered with the discovery of "new" stars (which were actually *supernovae*—exploding old stars). The impressive advances in physics and chemistry during the seventeenth and eighteenth centuries did point out, however, that some essences were indeed more basic and more permanent than others, and that a few were almost timeless for many practical purposes. For example, it was realized that chemical elements such as oxygen and carbon were constant (at least throughout human history) in their basic properties—the oxygen breathed by Julius Caesar was identical to that exhaled by Isaac Newton. Similarly, the laws of motion and of gravity formulated by Newton applied everywhere, from falling apples to the orbits of planets, and appeared to be positively unchangeable. However, in the absence of any clear guidelines as to how to determine which natural quantities or concepts were genuinely fundamental and which were not (in spite of some valiant efforts by empiricist philosophers such as John Locke, George Berkeley, and David Hume), many of the eighteenth-century naturalists opted to simply adopt the ancient Greek view of ideal, unchanged species.

These were the prevailing tides and currents of thought about life, until one man had the chutzpah, the vision, and the deep insights to weave together a huge set of separate clues into one magnificent tapestry. This man was Charles Darwin (figure 1 shows him late in life), and his grand unified conception has become humankind's

Figure 1

most inspiring nonmathematical theory. Darwin has literally transformed the ideas on life on Earth from a myth into a science.

Revolution

The first edition of Darwin's book *On the Origin of Species* was published on November 24, 1859, in London, and biology was changed forever on that day. (Figure 2 shows the title page of the first edition; Darwin referred to it as "my child" upon publication.) Before we examine the central arguments of *The Origin*, however, it is important to understand what is *not* discussed in that book. Darwin does not say even one word either about the actual *origin* of life or about the *evolution* of the universe as a whole. Furthermore, contrary to some popular beliefs, he also does not discuss at all the evolution of humans, except in one prophetic, optimistic paragraph near the end of the book, where he says, "In the distant future I see open fields for more important researches. Psychology will be based on a new foundation, that of the necessary acquirement of each mental power and capacity by graduation. Light will be thrown on the origin of man and his history." Only in a later book, *The Descent of Man and Selection in Relation to Sex,* which was published about a dozen years after *The Origin,* did Darwin decide to make it clear that he believed that his ideas on evolution should also apply to humans. He was actually much more specific than that, concluding that humans were the natural descendants of apelike creatures that probably lived in trees in the "Old World" (Africa):

> We thus learn that man is descended from a hairy, tailed quadruped, probably arboreal in its habits and an inhabitant of the Old World. This creature, if its whole structure had been examined by a naturalist, would have been classed among the Quadrumana [primates with four hands, such as apes], as surely as the still more ancient progenitor of the Old and New World monkeys.

Most of the intellectual heavy lifting on evolution, however, had already been achieved in *The Origin.* In one blow, Darwin disposed of the notion of design, dispelled the idea that species are eternal and immutable, and proposed a mechanism by which adaptation and diversity could be accomplished.

In simple terms, Darwin's theory consists of four main pillars that are supported by one remarkable mechanism. The pillars are: *evolution, gradualism, common descent,* and *speciation.* The crucial mechanism that drives it all and glues the different elements into cooperation is *natural selection,* which, we know today, is sup-

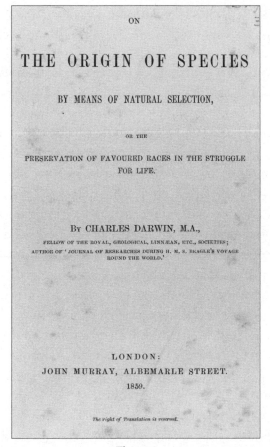

ON

THE ORIGIN OF SPECIES

BY MEANS OF NATURAL SELECTION,

OR THE

PRESERVATION OF FAVOURED RACES IN THE STRUGGLE
FOR LIFE.

By CHARLES DARWIN, M.A.,

FELLOW OF THE ROYAL, GEOLOGICAL, LINNÆAN, ETC., SOCIETIES;
AUTHOR OF ' JOURNAL OF RESEARCHES DURING H. M. S. BEAGLE'S VOYAGE
ROUND THE WORLD.'

LONDON:
JOHN MURRAY, ALBEMARLE STREET.
1859.

The right of Translation is reserved.

Figure 2

plemented to some degree by a few other vehicles of evolutionary change, some of which could not have been known to Darwin.

Here is a very succinct account of these distinct components of Darwin's theory. The description will mostly trace Darwin's own ideas rather than updated, modernized versions of these concepts. Still, in a few places, it will be essentially impossible to avoid the delineation of evidence that has accumulated since Darwin's time. As we shall discover in the next chapter, however, Darwin did make one serious error that could have negated entirely his most important insight: that of natural selection. The root of the error was not Darwin's fault—nobody in the nineteenth century understood genetics—but Darwin did not realize that the theory of genetics with which he was operating was lethal for the concept of natural selection.

The first essence in the theory was that of evolution itself. Even though some of Darwin's ideas on evolution had an older pedigree, the French and English naturalists that preceded him (among whom, figures such as Pierre-Louis Moreau de Maupertuis, Jean-Baptiste Lamarck, Robert Chambers, and Darwin's own grandfather, Erasmus Darwin, stood out) failed to provide a convincing mechanism for evolution to take place. Here is how Darwin himself described evolution: "The view which most naturalists entertain, and which I formerly entertained—namely, that each species has been independently created—is erroneous. I am fully convinced that species are not immutable; but that those belonging to what are called the same genera are lineal descendants of some other and generally extinct species." In other words, the species that we encounter today did not always exist. Rather, these are the descendants of some earlier species that became extinct. Modern biologists tend to distinguish between *microevolution* and *macroevolution*. Microevolution encompasses small changes (such as those sometimes observed in bacteria) that are the results of the evolutionary process over relatively short periods of time, typically within local populations. Macroevolution refers to the results of evolution over long timescales, typically among species—and which could also involve mass extinction episodes, such as the one that snuffed out the dinosaurs. In the years since the publication of *The Origin,* the idea of evolution has become so much the guiding principle of all the research in the life sciences that in 1973 Theodosius Dobzhansky, one of the twentieth century's most eminent evolutionary biologists, published an essay entitled "Nothing in Biology Makes Sense Except in the Light of Evolution." At the end of this article, Dobzhansky noted that the twentieth-century French philosopher and Jesuit priest Pierre Teilhard de Chardin "was a creationist, but one who understood that the Creation is realized in this world by means of evolution."

Darwin borrowed the idea embodied in his second pillar, that of gradualism, mainly from the works of two geologists. One was the eighteenth-century geologist James Hutton, and the other was Darwin's contemporary and later close friend Charles Lyell. The geological record showed horizontal banding patterns covering large geographical areas. This, coupled with the uncovering of different fossils within these bands, suggested a progression of incremental change. Hutton and Lyle were largely responsible for the formulation of the modern theory of *uniformitarianism:* the notion that the rates at which processes such as erosion and sedimentation occur at present are similar to the rates in the past. (We shall return to this concept in chapter 4, when we'll discuss Lord Kelvin.) Darwin argued that just as geological action shapes the Earth gradually but surely, evolutionary changes are the result of transformations that span hundreds of thousands of generations. One should not, therefore, expect to see significant alterations in less than tens of thousands of years, except perhaps in organisms that multiply very frequently, such as bacteria, which, as we know today, can develop resistance to antibiotics in extremely short times. Contrary to uniformitarianism, however, the rate of evolutionary changes is generally nonuniform in time for a given species, and it can vary further from one species to another. As we shall see later, it is the pressure exerted by natural selection that determines primarily how fast evolution manifests itself. Some "living fossils" such as the lamprey—a jawless marine vertebrate with a funnel-like mouth—appear to have hardly evolved in 360 million years. As a fascinating aside, I should note that the idea of gradual change was put forth in the seventeenth century by the empiricist philosopher John Locke, who wrote insightfully, "The boundaries of the species, whereby men sort them, are made by men."

The next pillar in Darwin's theory, the concept of a *common ancestor*, is what has become in its modern incarnation the primary motivator for all of the present-day searches for the origin of life. Darwin first argued that there is no doubt that all the members of any taxonomic class—such as all vertebrates—originated from a common ancestor. But his imagination carried him much further with this concept. Even though his theory predated any knowledge of the facts that all living organisms share such characteristics as the DNA molecule, a small number of amino acids, and the molecule that serves as the currency for energy production, Darwin was still bold enough to proclaim, "Analogy would lead me one step further, namely, to the belief that all animals and plants have descended from some one prototype." Then, after cautiously acknowledging that "analogy may be a deceitful guide," he still concluded that "probably all the organic beings which have ever lived on the earth have descended from some one primordial form, into which life was first breathed."

But, you may wonder, if all life on Earth originated from a single, common ancestor, how did the astonishing wealth of diversity arise? After all, this was the first hallmark of life that we have identified as one that requires an explanation. Darwin did not flinch, and took this challenge head-on—it was not an accident that the title of his book had the word "species" in it. Darwin's solution to the diversity problem involved another original idea: that of branching, or speciation. Life starts from a common ancestor, just as a tree has a single trunk, Darwin reasoned. In the same way that the trunk develops branches, which then split into twigs, the "tree of life" evolves by many branching and ramification events, creating separate species at each splitting node. Many of these species become extinct, just like the dead and broken branches of a tree. However, since at each splitting the number of offspring species from a given ancestor doubles, the number of different species can increase dramatically. When does speciation actually occur? According to modern thinking, mainly when a group of members of a particular species becomes geographically separated. For instance, one group may wander to the rainy side of a mountain range, while the rest of the species stays on the dry slope. Over time, these rather different environments produce different evolutionary paths, eventually leading to two populations that can no longer interbreed—or in other words, different species. In rarer occasions, speciation could create new species that arise from interbreeding between two species. Such appears to have been the case of the Italian sparrow, which was shown in 2011 to be genetically intermediate between Spanish sparrows and house sparrows. Italian and Spanish sparrows behave like distinct species, but Italian and house sparrows do form hybrid zones, where the ranges of the two interbreeding species meet.

Amazingly, in 1945, author Vladimir Nabokov, of *Lolita* and *Pale Fire* fame, came up with a sweeping hypothesis for the evolution of a group of butterflies known as the Polyommatus blues. Nabokov, who had a lifelong interest in butterflies, speculated that the butterflies came to the New World from Asia in a series of waves lasting millions of years. To their surprise, a team of scientists using gene-sequencing technology confirmed Nabokov's conjecture in 2011. They found that the New World species shared a common ancestor that lived about ten million years ago, but that many New World species were more closely related to Old World butterflies than to their neighbors.

Darwin was sufficiently aware of the importance of the concept of speciation to his theory to include a schematic diagram of his tree of life. (Figure 3 shows the original drawing from his 1837 notebook.) In fact, this is the only figure in the entire book. Fascinatingly, Darwin included the caveat "I think" at the top of the page!

In many cases, evolutionary biologists have been able to identify most of the intermediate steps involved in speciation: from pairs of species that have probably recently split from a single species, to pairs that are just about ready to be pushed into separation. At the more detailed level, a combination of molecular and fossil data has yielded, for instance, a relatively well-resolved and well-dated phylogenetic tree for all the families of living and very recently extinct mammals.

I cannot refrain at this point from digressing to note that from my own personal perspective, there is another aspect of the notions of a common ancestor and of speciation that makes Darwin's theory truly special. About a decade ago, while working on the book *The Accelerating Universe,* I was trying to identify the ingredients that make a physical theory of the universe "beautiful" in the eyes of scientists. In the end, I concluded that two of the absolutely essen-

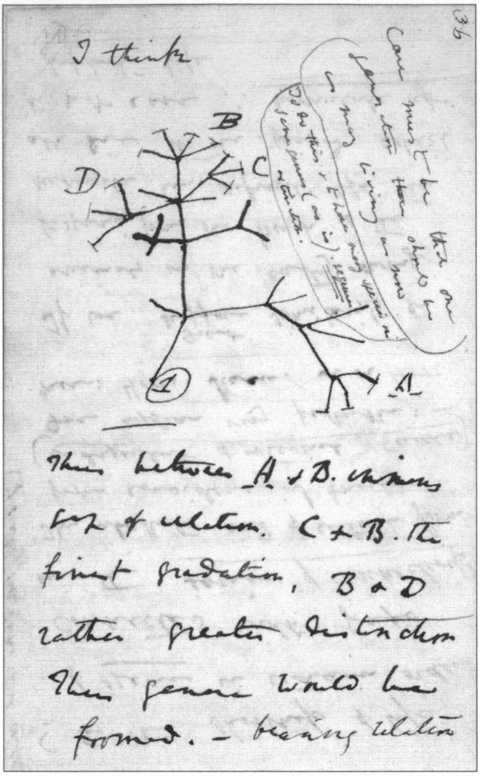

Figure 3

tial constituents were *simplicity* and something that is known as the *Copernican principle.* (In the case of physics, the third ingredient was *symmetry.*) By "simplicity," I mean reductionism, in the sense that most physicists understand it: the ability to explain as many phenomena as possible with as few laws as possible. This has always been, and still is, the goal of modern physics. Physicists are not satisfied, for instance, with having one extremely successful theory (quantum mechanics) for the subatomic world, and one equally successful theory (general relativity) for the universe at large. They would like to have one unified "theory of everything" that would explain it all.

The Copernican principle derives its name from that of the Polish astronomer Nicolaus Copernicus, who in the sixteenth century removed the Earth from its privileged position at the center of the universe. Theories that obey the Copernican principle do not require humans to occupy any special place for these theories to work. Copernicus taught us that the Earth is not at the center of the solar system, and all the subsequent findings in astronomy have only strengthened our realization that, from a physics perspective, humans play no special role in the cosmos. We live on a tiny planet that revolves around an ordinary star, in a galaxy that contains hundreds of billions of similar stars. Our physical insignificance continues even further. Not only are there about two hundred billion galaxies in our observable universe, but even ordinary matter—the stuff that we and all the stars and gas in all the galaxies are made of—constitutes only a little over 4 percent of the universe's energy budget. In other words, we are really nothing special. (In chapter 11 I will discuss some ideas suggesting that we should not take Copernican modesty too far.)

Both reductionism and the Copernican principle are the true trademarks of Darwin's theory of evolution. Darwin explained just about everything related to life on Earth (except its origin) with one unified vision. One can hardly be more reductionistic than that. At the same time, his theory was Copernican to the core. Humans evolved just like every other organism. In the tree analogy, all of the youngest buds are separated from the main trunk by a similar number of branching nodes, the only difference being that they point in different directions. Equivalently, in Darwin's evolutionary scheme, all the present-day living organisms, including humans, are the products of similar paths of evolution. Humans definitely do not occupy any exceptional or unique place in this scheme—they are not the lords of creation—but an adaptation and development of their ancestors on Earth. This was the end of "absolute anthropocentrism." All the terrestrial creatures are part of the same big family. In the words of the influential evolutionary biologist Stephen Jay Gould, "Darwinian evolution is a bush, not a ladder." To a large extent, what has fueled the opposition to Darwin for more than 150 years is precisely this fear that the theory of evolution displaces humans from the pedestal on which they have put themselves. Darwin has initiated a rethinking of the nature of the world and of humans. Note that in a picture in which only the "fittest" survive (as we shall soon discuss in the context of natural selection), one could argue that insects have clearly outclassed humans, since there are so many more of them. Indeed, the British geneticist J. B. S. Haldane is cited (possibly apocryphally) as having replied to theologians who inquired whether there was anything that could be concluded about the Creator from the study of creation, with the observation that God "has an inordinate fondness for beetles." Today we know that even in terms of genome size—the entirety of the hereditary information—humans fall far short of, believe it or not, a fresh water ameboid named *Polychaos dubium*. With 670 billion base pairs of DNA reported, the genome of this microorganism may be more than two hundred times larger than the human genome!

Darwin's theory, therefore, amply satisfies the two applicable criteria (which admittedly are somewhat subjective) for a truly beautiful theory. No wonder, then, that *The Origin* has elicited perhaps the most dramatic shift of thought ever brought about by a scientific treatise.

Returning now to the theory itself, Darwin was not content with merely making statements about evolutionary changes and the production of diversity. He regarded it as his main task to explain *how* these processes have occurred. To achieve this goal, he had to come up with a convincing alternative to creationism for the apparent design in nature. His idea—natural selection—has been esteemed by Tufts University philosopher Daniel C. Dennett as no less than "the single best idea anyone has ever had."

Natural Selection

One of the challenges that the concept of evolution posed concerned adaptation: the observation that species appeared to be perfectly harmonized with their environments, and the mutual adaptedness of the traits of organisms—body parts and physiological processes—to one another. This created a puzzle that confounded even the evolutionary minded among the naturalists that preceded Darwin: If species are so well adapted, how could they evolve and still remain well adapted? Darwin was fully aware of this conundrum, and he made sure that his principle of natural selection provided a satisfactory solution.

The basic idea underlying natural selection is quite simple (once it is pointed out!). As it sometimes happens with discoveries whose time has come, the naturalist Alfred Russel Wallace independently formulated very similar ideas at about the same time. Wallace was nevertheless very clear on who he thought deserved most of the credit. In a letter to Darwin on May 29, 1864, he wrote:

> As to the theory of Natural Selection itself, I shall always maintain it to be actually yours and yours only. You had worked it out in details I had never thought of, years before I had a ray of light on the subject, and my paper would never have convinced anybody or been noticed as more than an ingenious speculation, whereas your book has revolutionized the study of Natural History.

Let us attempt to follow Darwin's train of thought: First, he noted, species tend to produce more offspring than can possibly survive. Second, the individuals within a given species are never all precisely identical. If some of them possess any kind of advantage in terms of their ability to cope with the adversity of the environment—and *assuming that this advantage is heritable, and passed on to their descendants*—then over time, the population will gradually shift toward organisms that are better adapted. Here is how Darwin himself put it, in chapter 3 of *The Origin*:

> Owing to this struggle for life, any variation, however slight and from whatever cause proceeding, if it be in any degree profitable to an individual of any species, in its infinitely complex relations to other organic beings and to

external nature, will tend to the preservation of that individual, and will generally be inherited by its offspring. The offspring, also, will thus have a better chance of surviving, for, of the many individuals of any species which are periodically born, but a small number can survive. I have called this principle, by which each slight variation, if useful, is preserved, by the term of Natural Selection.

Using the modern gene terminology (of which Darwin knew absolutely nothing), we would say that natural selection is simply the statement that those individuals whose genes are "better" (in terms of survival and reproduction) would be able to produce more offspring, and that those offspring will also have better genes (relatively speaking). In other words, over the course of many generations, beneficial mutations will prevail, with harmful ones eliminated, resulting in evolution toward better adaptation. For instance, it is easy to see how being faster could benefit both predator and prey. So in East Africa's open plains of the Serengeti, natural selection has produced some of the fastest animals on Earth.

There are several elements that combine effectively to create the complete picture of natural selection. First, natural selection takes place in *populations*—communities of interbreeding individuals at given geographical locations—not in individuals. Second, populations typically have such high reproduction potential that if unchecked they would increase exponentially. For example, the female of the ocean sunfish, Mola mola, produces as many as three hundred million eggs at a time. If even just 1 percent of those eggs are fertilized and survive to adulthood, we soon would have oceans filled with Mola molas (and the average weight of an adult ocean sunfish exceeds two thousand pounds). Fortunately, due to competition for resources within the species, struggles with predators, and the environment's other adversities, from a set of parents belonging to any species, an average of only two offspring survive and reproduce.

This description makes it clear that the word "selection" in Darwin's formulation of natural selection really refers more to a process of *elimination* of the "weaker" (in terms of survival and reproduction) members of a population, rather than to a selection by an anthropomorphic nature. Metaphorically, you could think of the process of selection as one of sifting through a giant sieve. The larger particles (corresponding to those that survive) remain in the sieve, while the ones that pass through are eliminated. The environment is the agent that does the shaking of the sieve. Consequently, in a letter that Wallace wrote to Darwin on July 2, 1866, he actually suggested that Darwin should consider changing the name of the principle:

> I wish, therefore, to suggest to you the possibility of entirely avoiding this source of misconception . . . and I think it may be done without difficulty and very effectually by adopting Spencer's term (which he generally uses in preference to Natural Selection), viz. "Survival of the Fittest." This term is the plain expression of the facts; "Natural Selection" is a metaphorical expression of it, and to a certain degree indirect and incorrect, since, even personifying Nature, she does not so much select special variations as exterminate the most unfavourable ones.

Darwin adopted this expression, coined in 1864 by the polymath Herbert Spencer, as a synonym for natural selection in his fifth edition of *The Origin*. However, present-day biologists rarely use this term, since it may give the wrong impression that it means that only the strong or healthy survive. In fact, "survival of the fittest" meant to Darwin precisely the same as "natural selection." That is, those organisms with *selectively favored and heritable characteristics* are the ones who most successfully pass those to their offspring. In this sense, even though Darwin admitted to having been inspired by ideas of philosophical radicals such as the political economist Thomas Malthus—some sort of biological economics in a world of free competition—important differences exist.

A third and extremely important point to note about natural selection is that it really consists of two sequential steps, the first of which involves primarily randomness or chance, while the second one is definitely nonrandom. In the first step, a heritable *variation* is produced. In modern biological language, we understand this to be a genetic variation introduced by random mutations, gene reshuffling, and all the processes associated with sexual reproduction and the creation of a fertilized egg. In the second step, *selection*, those individuals in the population that are best suited to compete, be it with members within their own species, with members of other species, or in terms of their ability to cope with the environment, are more likely to survive and reproduce. Contrary to some misconceptions about natural selection, chance plays a much smaller role in the second step. Nevertheless, the process of selection is still not entirely deterministic—good genes are not going to help a species of dinosaurs wiped out by the impact of a giant meteorite, for instance. In a nutshell, therefore, evolution is really a change over time in the frequency of genes.

There are two main features that distinguish natural selection from the concept of "design." First, natural selection does not have any long-term "strategic plan" or ultimate goal. (It is not teleological.) Rather than striving toward some ideal of perfection, it simply tinkers by elimination of the less adapted with generation after generation, often changing direction or even resulting in the extinction of entire lineages. This is not what one would expect from a master designer. Second, because natural selection is constrained to work with what already exists, there is only so much that it can actually achieve. Natural selection starts by modifying species that have already evolved to a certain state, rather than by redesigning them from scratch. This is similar to asking a tailor to do some alterations to an old dress instead of asking the Versace fashion house to design a new one. Consequently, natural selection leaves quite a bit to be desired in terms of design. (Wouldn't a visual field covering all 360 degrees or having four hands be nice? And were having nerves in the teeth or a prostate gland that totally surrounds the urethra really such great ideas?) So even if certain characteristics confer a fitness advantage, as long as there is no heritable variation that achieves this result, natural selection could never produce such characteristics. Imperfections are, in fact, natural selection's unmistakable fingerprint.

You have probably noticed that Darwin's theory of evolution is, by its very nature, not easily provable by direct evidence, since it typically operates on such long timescales that watching grass grow feels like a fast-paced action movie by comparison. Darwin himself wrote to the geologist Frederick Wollaston Hutton on April 20, 1861, "I am actually weary of telling people that I do not pretend to adduce evidence of one species turning into another, but I believe that this view is in the main correct, because so many phenomena can thus be grouped and explained." Nevertheless, biologists, geologists, and paleontologists have amassed a huge body of circumstantial evidence for evolution, most of which is beyond the scope of this book, since it is not related directly to Darwin's blunder. Let me only note the following fact: The fossil record reveals an unmistakable evolution from simple to complex life. Specifically, over the billions of years of geological time, the more ancient the geological layer in which a fossil is uncovered, the simpler the species.

It is important to mention briefly a few of the pieces of evidence supporting the idea of natural selection, since it was the notion that life could evolve and diversify without there being a goal to evolve *toward* that was the most deeply unsettling aspect of the theory to Darwin's contemporaries. I have already mentioned one clue demonstrating the reality of natural selection: the resistance to drugs developed by various pathogens. The bacterium known as *Staphylococcus aureus,* for instance, is the most common cause for the types of infections known as staph infections, which affect no fewer than a half million patients in American hospitals each year. In the early 1940s, all the known strains of staph were susceptible to penicillin. Over the years, however, due to mutations producing resistance and through natural selection, most staph strains have become resistant to penicillin. In this case, the entire process of evolution has been compressed in time dramatically (due partly to the selective pressure exerted by humans), since the generations of bacteria are so short lived and the population is so enormous. Since 1961, a particular staph strain known as MRSA (an acronym for methicillin-resistant *Staphylococcus aureus*) has developed resistance not just to penicillin but also to methicillin, amoxicillin, oxacillin, and a whole host of other antibiotics. There is hardly a better manifestation of natural selection in action.

Another fascinating (although controversial) example of natural selection is the evolution of the peppered moth. Prior to the industrial revolution, the light colors of this moth (known among biologists as *Biston betularia betularia morpha typica*) provided ample camouflage against the background of its habitat: lichens and trees. The industrial revolution in England brought with it immense levels of pollution that destroyed many lichens and blackened many trees with soot. Consequently, the white-bodied moths were exposed suddenly to massive predation, which led to their near extinction. At the same time, the melanic, dark-colored variety of the moth (*carbonaria*) started to flourish around 1848, because of its much improved camouflage characteristics. As if to demonstrate the importance of "green" practices, the white-bodied moths started reappearing again once better environmental standards had been adopted. While some studies of the peppered moth and the phenomenon described above ("industrial melanism") have been criticized by a number of creationists, even some of the critics agree that this is a clear case of natural selection, and they argue only that this does not provide proof of evolution, since the net result is merely of one type of moth morphing into another rather than into an entirely new species altogether.

Another common, more philosophical, objection to natural selection is that Darwin's definition of it is circular, or *tautological.* Put in simple terms, the adverse judgment goes something like this: Natural selection means "survival of the fittest." But how do you define the "fittest"? They are identified as those that survive best; hence, the definition is a tautology. This argument stems from a misunderstanding, and it is absolutely false. Darwin did not use "fitness" to refer to those who survive but to those who, when compared with other members of the species, could be *expected* to survive *because they were better adapted to the environment.* The interaction between a variable feature of an organism and the environment of that organism is crucial here. Since the organisms compete for limited resources, some survive and some don't. Furthermore, for natural selection to operate, the adaptive characteristics need to be *heritable*, that is, capable of being genetically passed on.

Surprisingly, even the famous philosopher of science Karl Popper raised a suspicion of tautology against evolution by natural selection (albeit a more subtle one). Popper basically questioned natural selection's explanatory power based on the following argument: If certain species exist, this means that they were adapted to their environment (since those that were not adapted became extinct). In other words, Popper asserted, adaptation is simply *defined* as the quality that guarantees existence, and nothing is ruled out. However, since Popper published this argument, a number of philosophers have shown it to be erroneous. In reality, Darwin's theory of evolution rules out more scenarios than it leaves in. According to Darwin, for instance, no new species can emerge without having an ancestral species. Similarly, in Darwin's theory, any variations that are not achievable in gradual steps are ruled out. In modern terminology, "achievable" would refer to processes governed by the laws of molecular biology and genetics. A key point here is the statistical nature of adaptation—no predictions can be made about individuals, just about probabilities. Two identical twins are not guaranteed to produce the same number of offspring, or even to both survive. Popper, by the way, did recognize his error in later years, declaring, "I have changed my mind about the testability and the logical status of natural selection; and I am glad to have an opportunity to make a recantation."

Finally, for completeness, I should mention that although natural selection is the main driver of evolution, other processes can bring about evolutionary changes. One example (which Darwin could not have known about) is provided by what has been termed by modern evolutionary biologists *genetic drift:* a change in the relative frequency in which a variant of a gene (an *allele*) appears in a population due to chance or sampling errors. This effect can be significant in small populations, as the following examples demonstrate. When you flip a coin, the expectation is that heads will turn up about 50 percent of the time. This means that if you flip a coin a million times, the number of times you'll get heads will be close to a half million. If you toss a coin just four times, however, there is a nonnegligible probability (of about 6.2 percent) that it will land heads each time, thus deviating substantially from the expectation. Now imagine a very large island population of organisms in which just one gene appears in two variants (alleles): X or Z. The alleles have an equal frequency in the population; that is, the frequency of X and Z is ½ for each. Before these organisms have a chance to reproduce, however, a huge tsunami wave washes the island, killing all but four of the organisms. The surviving four organisms could have any of the following sixteen combinations of alleles: XXXX, XXXZ, XXZX, XZXX, ZXXX, XXZZ, ZZXX, XZZX, ZXXZ, XZXZ, ZXZX, XZZZ, ZZZX, ZXZZ, ZZXZ, ZZZZ. You will notice that in ten out of these sixteen combinations, the number of X alleles is *not* equal to the number of Z alleles. In other words, in the surviving population, there is a higher chance for a genetic drift—a change in the relative allele frequency—than for keeping the initial state of equal frequencies.

Genetic drift can cause a relatively rapid evolution in a small population's gene pool, which is independent of natural selection. One oft-cited example of genetic drift involves the Amish community of eastern Pennsylvania. Among the Amish, polydactyly (extra fingers or toes) is many times more common than in the general population of the United States. This is one of the manifestations of the rare Ellis-van Creveld syndrome. Diseases of recessive genes, such as the Ellis-van Creveld syndrome, require two copies of the gene to cause the disease. That is, both parents have to be carriers of the recessive gene. The reason for the higher-than-normal frequency of these genes in the Amish community is that the Amish marry within their own group, and the population itself originated from around two hundred German immigrants. The small size of this community allowed researchers to trace back the Ellis-van Creveld syndrome to just one couple, Samuel King and his wife, who arrived in 1744.

There are three points that need to be emphasized about genetic drift. First, the evolutionary changes that are due to genetic drift occur entirely as a result of chance and sampling errors—they are not driven by selection pressure. Second, genetic drift cannot cause adaptation, which remains entirely the province of natural selection. In fact, being entirely random, genetic drift can cause certain properties to evolve whose usefulness is otherwise very puzzling. Finally, while genetic drift clearly occurs to some degree in all populations (since all the populations are finite in size), its effects are most pronounced in small, isolated populations.

These are, very concisely, some of the key points of Darwin's theory of evolution by natural selection. Darwin revolutionized biological thinking in two major ways. He not only recognized that beliefs held for centuries could be false but also demonstrated that scientific truth can be achieved by the patient collection of facts, coupled with bold hypothesizing about the theory that binds those facts together. As you must have realized, his theory does a superb job in explaining why life on Earth is so diverse and why living organisms have the characteristics they have. The nineteenth-century English suffragist and botanist Lydia Becker beautifully described Darwin's achievement:

> How seemingly unimportant are the movements of insects, creeping in and out of flowers in search of the nectar on which they feed! If we saw a man spending his time in watching them, and in noting their flitting with curious eyes, we might be excused for imagining that he was amusing himself by idling an hour luxuriously in observing things which, though curious, were trifling. But how mistaken might we be in such an assumption! For these little winged messengers bear to the mind of the philosophical naturalist tidings of mysteries hitherto unrevealed; and as Newton saw the law of gravitation in the fall of the apple, Darwin found, in the connection between flies and flowers, some of the most important facts which support the

theory he has promulgated respecting the modification of specific forms in animated beings.

Indeed, Darwin was to the nineteenth century what Newton was to the seventeenth, and Einstein to the twentieth. It is curious that the theory of *evolution* constituted one of the most dramatic *revolutions* in the history of science. In the words of biologist and science historian Ernst Mayr, it "caused a greater upheaval in man's thinking than any other scientific advance since the rebirth of science in the Renaissance." The question, then, is: Where was Darwin's blunder?

CHAPTER 3

YEA, ALL WHICH IT INHERIT, SHALL DISSOLVE

Life's perhaps the only riddle
That we shrink from giving up!

—WILLIAM SCHWENCK GILBERT,
THE GONDOLIERS

The title of this chapter is taken partly from William Shakespeare's *The Tempest*, but as we shall soon see, it poetically captures the essence of Darwin's blunder. The source of the blunder was the fact that the prevailing theory of heredity in the nineteenth century was fundamentally flawed. Darwin himself was aware of the existing shortcomings, as he confessed candidly in *The Origin:*

> The laws governing inheritance are quite unknown; no one can say why the same peculiarity in different individuals of the same species, and in individuals of different species, is sometimes inherited and sometimes not so; why the child often reverts in certain characters to its grandfather or grandmother or other much more remote ancestor; why a peculiarity is often transmitted from one sex to both sexes, or to one sex alone, more commonly but not exclusively to the like sex.

To say that the laws of inheritance were "quite unknown" was probably the most glaring understatement of the entire book. Darwin had been educated according to the then widely held belief that the characteristics of the two parents become physically blended in their offspring—as in the mixing of paints. In this "paint-pot theory," the heredity contribution of each ancestor was predicted to be halved in each generation, and the offspring of any sexual partners were expected to be intermediates. In Darwin's own words: "After twelve generations, the proportion of blood, to use a common expression, of any one ancestor is only 1 in 2,048." That is, as with gin and tonic, if you keep mixing the drink with tonic, you eventually no longer taste the gin. Somehow, in spite of apparently understanding this inevitable dilution, Darwin still expected natural selection to work. For instance, in his example of wolves preying on deer, he concluded, "If any slight innate change of habit or of structure benefited an individual wolf, it would have the best chance of surviving and leaving progeny. Some of its young would probably inherit the same habits or structure, and by the repetition of this process, a new variety might be formed." But the simple fact that this expectation was absolutely untenable under the assumption of a blending theory of heredity did not occur to Darwin. The inconsistency was first noted by the Scottish engineer Fleeming Jenkin.

Jenkin was a multitalented individual whose pursuits ranged from drawing portraits of passersby to designing undersea telegraph cables. His criticism of Darwin was fairly straightforward. Jenkin argued that natural selection would be totally ineffective in "selecting" a *single variation* (a rare novelty that arose by chance, which he referred to as a "sport"; today we would call it a mutation), because any such variation would be *swamped* and diluted by all the normal types in the population and obliterated entirely after a few generations.

Darwin could not be faulted for not knowing any better than the heredity theory that was scientifically accepted at his time. Consequently, I do not consider his adopting the idea of blending inheritance as a blunder. Darwin blundered in *having completely missed the point (at least initially) that his mechanism of natural selection simply could not work as envisioned, under the assumption of blending inheritance.* Let us examine this serious blunder and its potentially devastating consequences in more detail.

Swamping

Fleeming Jenkin published his criticism of Darwin's theory as an anonymous review of the fourth edition of *On the Origin of Species.* The article appeared in the *North British Review* in June 1867. While the essay attacked the theory of evolution on several grounds, I shall concentrate here on the one argument that exposed Darwin's blunder. To demonstrate his point, Jenkin assumed that each individual has one hundred offspring, but of those, on the average, only one survives to reproduce. He then discussed an individual with a rare mutation ("sport") that has the advantage of having twice the chance of survival and reproduction as any other. Appropriately for the rigorous engineer that he was (he received no fewer than thirty-seven patents between 1860 and 1886), Jenkin's approach was quantitative—he wanted to actually calculate the effect of such a "sport" on the general population:

> It will breed and have a progeny of say 100; now this progeny will, on the whole, be intermediate between the average individual and the sport. [Since the sports are rare, a sport is expected to mate with an average individual.] The odds in favour of one of this generation of the new breed will be, say, 1.5 to 1 [under the assumption of blending], as compared with the average individual; the odds in their favor will therefore be less than that of their parent; but owing to their greater number, the chances are that about 1.5 of them would survive. Unless these breed together, a most improbable event, their progeny would again approach the average individual; there would be 150 of them [1.5 times 100], and their superiority would be say in the ratio of 1.25 to 1

[again because of blending]; the probability would now be that nearly two of them would survive [1 percent of 1.25 times 150] and have 200 children, with an eighth superiority. Rather more than two of these would survive; but the superiority would again dwindle, until after a few generations it would no longer be observed, and would count no more in the struggle for life, than any of the hundred trifling advantages which occur in the ordinary organs.

Jenkin argued that even under the most extreme form of selection, one could not expect the complete transformation of a well-established characteristic, such as skin color, into a new one, if that new characteristic had been introduced into the population only once. To illustrate this swamping effect, Jenkin chose a startlingly prejudicial example of a white man with superior characteristics shipwrecked on an island inhabited by blacks. The racist and imperialistic tone of the passage utterly shocks us today, but it probably was commonplace in late-Victorian Britain: Even if this person "would kill many blacks in the struggle for existence" and "would have a great many wives and children," and "in the first generation there will be some dozens of intelligent young mulattoes," Jenkin argued, "can any one believe that the whole island will gradually acquire a white, or even a yellow population?"

As it turned out, Jenkin actually made one serious logical mistake in his calculations. He assumed that each sexual pair had one hundred offspring, of whom, on the average, only *one* offspring survived to reproduce. However, since only females can reproduce, it follows that out of each mating couple, *two* offspring must on the average survive (one male and one female); otherwise the size of the population would be halved in each generation—a recipe for rapid extinction. Surprisingly, only Arthur Sladen Davis, an assistant mathematics master at Leeds Grammar School, discovered this obvious error, and he explained it in a letter to the journal *Nature* in 1871.

Davis showed that when a correction is made to keep the population roughly constant in size, the effect of a sport does not die out (as Jenkin contended), but, in fact, although diluted, it becomes distributed throughout the entire population. For instance, a black cat introduced into a population of white cats would (under the assumption of blending inheritance) on the average produce two gray kittens, four lighter grandkittens, and so on. Successive generations would become progressively lighter, but the dark hue would never disappear. Davis also concluded correctly that "though any favourable sport occurring once, and never again, except by inheritance, will effect scarcely any change in a race, yet that sport, arising independently in different generations, though never more than once in any one generation, may effect a very considerable change."

In spite of Jenkin's mathematical error, his general criticism was correct: On the supposition of blending inheritance, even under the most favorable conditions, a black cat occurring once could not turn an entire population of white cats black, no matter how advantageous the black color might have been.

Before we scrutinize the question of how Darwin could have missed this seemingly fatal shortcoming of his theory of natural selection, it would be helpful to understand the blending theory of heredity from the perspective of modern genetics.

Darwin's Blunder and the Seeds of Genetics

In the context of our current understanding of genetics, the molecule known as *DNA* (*deoxyribonucleic acid*) provides the mechanism responsible for heredity in all living organisms. Very roughly speaking, DNA is made up of *genes*, which contain the information that codes for proteins, and of some noncoding regions. Physically, DNA is located on elements called *chromosomes*, of which each individual organism in sexual species has two sets, one inherited from the mother (the female) and one from the father (the male). Consequently, each individual has two sets of all of its genes, where the two copies of a gene may be identical, or slightly different. The different forms of a gene that can be present at a particular location on a chromosome are the variants referred to as alleles.

The modern theory of genetics originated from the mind of an unlikely explorer: a nineteenth-century Moravian priest named Gregor Mendel. He performed a series of seemingly simple experiments in which he cross-pollinated thousands of pea plants that produce only green seeds with plants that produce only yellow seeds. To his surprise, the first offspring generation had only yellow seeds. The next generation, however, had a 3:1 ratio of yellow to green seeds. From these puzzling results, Mendel was able to distill a *particulate,* or *atomistic,* theory of heredity. In categorical contrast to blending, Mendel's theory states that genes (which he called "factors") are discrete entities that are not only preserved during development but also passed on *absolutely unchanged* to the next generation. Mendel further added that every offspring inherits one such gene ("factor") from each parent, and that a given characteristic may not manifest itself in an offspring but can still be passed on to the following generations. These deductions, like Mendel's experiments themselves, were nothing short of brilliant. Nobody had reached similar conclusions in almost ten thousand years of agriculture. Mendel's results at once disposed of the notion of blending, since already in the very first offspring generation, all the seeds were not an average of the two parents.

A simple example will help to clarify the key differences between Mendelian and blending heredity, in terms of their effects on natural selection. Even though blending inheritance clearly never used the concept of genes, we can still employ this language while preserving the essence of the process of blending. Imagine that organisms that carry a particular gene A are black, while the bearers of gene a are white. We will start with two individuals, one black and one white, each one having two copies of the respective gene (as in figure 4). If no gene dominates over the other, then in both blending heredity and Mendelian heredity, the offspring from such a couple would be gray, since they would have the gene combination (or *genotype*) Aa. Now, however, comes the key difference. In the blending theory, the A and the a would physically blend to create a new type of gene that gives its carrier the color gray. We can call this new gene $A^{(1)}$. Such

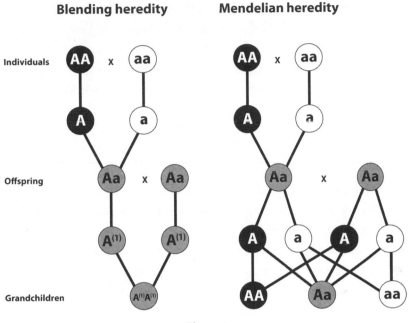

Figure 4

blending would not occur in Mendelian heredity, where each gene would keep its identity. As figure 4 shows, in the grandchildren's generation, all the offspring would be gray under blending heredity, while they could be black (*AA*), white (*aa*), or gray (*Aa*) under Mendelian heredity. In other words, Mendelian genetics pass down extreme genetic types from one generation to the next, thereby efficiently maintaining genetic variation. In blending heredity, on the other hand, variation is inevitably lost, as all the extreme types vanish rapidly into some intermediate mean. As Jenkin observed correctly, and the following (highly simplified) example will clearly demonstrate, this feature of blending heredity was catastrophic for Darwin's ideas on natural selection.

Imagine that we start with a population of ten individuals. Nine have the gene combination *aa* (and are therefore white), and one has the combination *Aa* (say, by some mutation), which renders it gray. Suppose further that being black is advantageous in terms of survival and reproduction, and that even having a somewhat darker color is better than being entirely white (although the advantage decreases with decreasing darkness). Figure 5 attempts to follow schematically the evolution of such a population under blending heredity. In the first generation, the blending of A with a will produce the new "gene" $A^{(1)}$, which, when mating with aa will yield $A^{(1)}a$, which will blend again to produce the gene $A^{(2)}$, corresponding to an even lighter and less advantageous color. You can easily see that after a large number (n) of generations, the most that can happen is that the population will be transformed into one with the combinations $A^{(n)}$ $A^{(n)}$, which will be only slightly darker than the original white population. In particular, the color black will become extinct even after the first generation, since its gene will be blended out of existence.

But under Mendelian heredity (figure 6), since the A gene is preserved from one generation to the next, eventually two Aa's will mate and produce the black AA variety. If black confers an advantage in the environment, then given enough time, natural selection could even turn the entire population black.

The conclusion is simple: For Darwin's theory of evolution by natural selection to really work, it needed Mendelian heredity. But in the absence of yet-undiscovered genetics, how did Darwin respond to Jenkin's criticism?

What Doesn't Kill You Makes You Stronger

Darwin was a genius in many ways, but he definitely was not a sharp mathematician. In his autobiography, he acknowledged, "I attempted mathematics, and even went during the summer of 1828 with a private tutor (a very dull man) to Barmouth, but I got on very slowly. The work was repugnant to me, chiefly from my not being able to see any meaning in the early steps of algebra . . . I do not believe that I should ever have succeeded beyond a very low grade." That being the case, arguments in *The Origin* are generally qualitative rather than quantitative, especially when it comes to the production of evolutionary change. In the few places where Darwin attempted to do simple calculations in *The Origin*, he managed occasionally to botch them.

Blending heredity

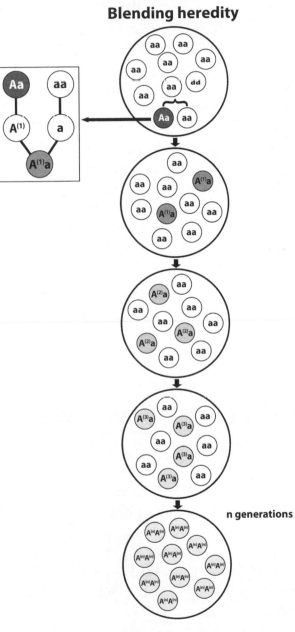

Figure 5

Mendelian heredity

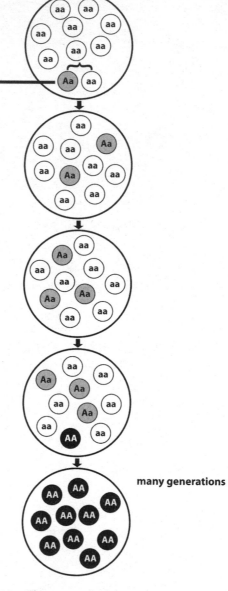

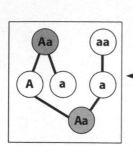

many generations

Figure 6

No wonder, then, that in one of his letters to Wallace, after reading Jenkin's rather mathematical criticism, he confessed, "I was blind and thought that single variations might be preserved much oftener than I now see is possible or probable." Still, it would have been amazing to think that Darwin had been totally unaware of the potential swamping effect of blending heredity until he read Jenkin's article. And indeed he wasn't. As early as 1842, twenty-five years before the publication of Jenkin's review, Darwin had already observed, "If in any country or district all animals of one species be allowed freely to cross, any small tendency in them to vary will be constantly counteracted." In reality, Darwin even relied to some extent on swamping to produce populational integrity in the face of the tendency of individuals to depart from their type due to variations. How did he then fail to understand how difficult it would be for a "sport" (a single variation) to fight off the averaging force of blending? Darwin's blunder and his slowness to recognize the point raised by Jenkin probably reflected on one hand his conceptual difficulties with heredity in general, and on the other, his residual overattachment to the idea that variations had to be scarce. The latter may have partially been a consequence of his general theory of reproduction and development, in which he assumed that only developmental stress triggered variations. Darwin's bafflement with heredity ran much deeper, as can be seen from the following inconsistency. At one point in *The Origin*, Darwin noted:

> When a character which has been lost in a breed, reappears after a great number of generations, the most probable hypothesis is, not that the offspring suddenly take after an ancestor some hundred generations distant, but that in each successive generation there has been a tendency to reproduce the character in question, which at last, under unknown favourable conditions, gains an ascendancy.

This notion of some latent "tendency" departed manifestly from normal blending heredity, and in many ways it was close in spirit to Mendelian heredity. Yet it apparently did not occur to Darwin, at least initially, to invoke this idea of latency in his struggle to respond to Jenkin. Instead, Darwin decided to change the emphasis from the role he had previously assigned to single variations to that of *individual differences* (the wide spectrum of tiny differences occurring frequently, which was supposed to be distributed continuously throughout the population), in supplying the "raw materials" for natural selection to effect. In other words, Darwin now relied on an entire continuum of variations for the production of evolution by natural selection over many generations.

In a letter to Wallace on January 22, 1869, the distressed Darwin wrote, "I have been interrupted in my regular work in preparing a new edition of the 'Origin,' which has cost me much labour, and which I hope I have considerably improved in two or three important points. I always thought individual differences more important than single variations, but now I have come to the conclusion that they [individual differences] are of paramount importance, and in this I believe I agree with you. Fleeming Jenkin's arguments have convinced me." To reflect his new emphasis, Darwin amended the fifth edition and subsequent editions of *The Origin* by changing singulars referring to individuals into plurals, as in "any variation" turning into "variations," and "an individual" into "individual differences." He also added a few new paragraphs in the fifth edition, two of which, in particular, are of great interest. In one, he admitted openly:

> I saw, also, that the preservation in a state of nature of any occasional deviation of structure, such as a monstrosity, would be a rare event; and that, if preserved, it would generally be lost by subsequent intercrossing with ordinary individuals. Nevertheless, until reading an able and valuable article in the "North British Review" (1867), I did not appreciate how rarely single variations, whether slight or strongly marked, could be perpetrated.

In the other paragraph, Darwin presented his own brief summary of Jenkin's swamping argument. This paragraph is fascinating because of two apparently small yet extremely significant differences from Jenkin's original text. First, Darwin assumes here that a pair of animals has two hundred offspring, of which *two* survive to reproduce. In spite of his nonmathematical background, therefore, Darwin appears to have anticipated already in 1869 the correction to Jenkin pointed out in A. S. Davis's letter to *Nature* in 1871: For the population not to disappear, two offspring, on the average, must survive. Second, and even more intriguing, Darwin assumes in his summary that only half of the offspring of the "sport" inherit the favorable variation. Note, however, that this assumption is contrary to the predictions of blending heredity! Unfortunately, Darwin was still unable at that time to elaborate on the possible consequences of a nonblending theory of heredity, and he accepted Jenkin's conclusions without any further discussion.

There are, nevertheless, quite a few signs that Darwin had not been happy with blending heredity for quite a while. In a letter he wrote in 1857 to the biologist Thomas Henry Huxley, his friend and champion in the public arena, he explained:

> Approaching the subject [of evolution] from the side which attracts me most, viz inheritance, I have lately been inclined to speculate very crudely and indistinctly, that propagation by true fertilization, will turn out to be a sort of mixture and not true fusion, of two distinct individuals, or rather innumerable individuals, as each parent has its parents and ancestors. I can understand on no other view the way in which crossed forms go back to so large an extent to ancestral forms. But all this, of course, is infinitely crude.

Crude or not, this observation was extremely insightful. Darwin recognized here that the combination of paternal and maternal heredity material was more like the shuffling together of two packs of cards rather than like the mixing of paints.

While Darwin's ideas in this letter can definitely be considered impressive forerunners of Mendelian genetics, Darwin was eventually driven by his frustration with blending heredity to develop a completely wrong theory known as *pangenesis.* In Darwin's pangenesis, the entire body was supposed to issue instructions to the reproductive cells. "I assume," he wrote in his book *The Variation of Animals and Plants Under Domestication,*

> that cells, before their conversion into completely passive or "formed material" throw off minute granules or atoms, which circulate freely through the system, and when supplied with proper nutriment multiply by self-division, subsequently becoming developed into cells like those from which they were derived. . . . Hence, speaking strictly, it is not the reproductive elements . . . which generate new organisms, but the cells themselves throughout the body.

To Darwin, the great advantage that pangenesis offered over blending was that if some adaptive change were to occur during the lifetime of an organism, then the granules (or "gemmules," as he called them) could take note of the change, lodge in the reproductive organs, and ensure that the change would be transmitted to the next generation. Unfortunately, pangenesis was taking heredity precisely in the opposite direction from which modern genetics was about to direct it—it is the fertilized egg that instructs the development of the entire body, not the other way around. Confused, Darwin clung to this misguided theory with similar conviction to that which he exhibited when he had previously held on to his correct theory of natural selection. In spite of vehement attacks by the scientific community, Darwin wrote to his great supporter Joseph Dalton Hooker in 1868: "I fully believe that each cell does actually throw off an atom or gemmule of its contents; but whether or not, this hypothesis serves as a useful connecting link for various grand classes of physiological facts, which at present stand absolutely isolated." He also added with confidence that even "if pangenesis is now stillborn, it will, thank God, at some future time reappear, begotten by some

other father, and christened by some other name." This was a perfect
example of a brilliant idea—particulate inheritance—that failed mis-
erably because it had been incorporated into the wrong mechanism
for its implementation: pangenesis.

Nowhere did Darwin articulate more clearly his atomistic, essentially Mendelian, ideas of heredity than in an exchange with Wallace
in 1866. First, in a letter written on January 22, he noted, "I know of
a good many varieties, which must be so called, that will not blend
or intermix, but produce offspring quite like either parent." Failing
to see Darwin's point, Wallace replied on February 4, "If you 'know
varieties that will not blend or intermix, but produce offspring quite
like either parent,' is not that the very physiological test of a species
which is wanting for the *complete proof* of the 'origin of species.'"

Realizing the misunderstanding, Darwin was quick to correct
Wallace in his next letter:

> I do not think you understand what I mean by the non-
> blending of certain varieties. It does not refer to fertility.
> An instance will explain. I crossed the Painted Lady and
> Purple sweet peas, which are very differently coloured
> varieties, and got, even out of the same pod, both varieties
> perfect, but none intermediate. Something of this kind,
> I should think, must occur at first with your butterflies
> and the three forms of Lythrum; though these cases are
> in appearance so wonderful, I do not know that they are
> really more so than every female in the world producing
> distinct male and female offspring.

This letter is remarkable in two ways. First, Darwin describes here
the results of experiments similar to those conducted by Mendel—
actually, the very experiments that had led Mendel to the formulation of Mendelian heredity. Darwin came pretty close to discovering
the Mendelian 3:1 ratio by himself. After he crossed the common
snapdragon (having bilateral symmetry) with the peloric (star-

shaped) form, the first generation of offspring were all of the common type, and the second had eighty-eight common to thirty-seven peloric (a ratio of 2.4:1). Second, Darwin points out the obvious fact that the simple observation that all offspring are either male or female, rather than some intermediate hermaphrodite, in itself argues against "paint-pot" blending! So the evidence of the proper form of heredity was right there in front of Darwin's eyes. As he had already remarked in *The Origin:* "The slight degree of variability in hybrids from the first cross or in the first generation, in contrast with their extreme variability in the succeeding generations, is a curious fact and deserves attention." Note also that the entire Darwin-Wallace correspondence above took place *before* the publication of Jenkin's review. All the same, even though Darwin came tantalizingly close to Mendel's discovery, he did not grasp its all-encompassing generality, and he failed to recognize its vital importance for natural selection.

To fully understand Darwin's attitude toward particulate heredity, there are a few other nagging questions that need to be resolved. Gregor Mendel read the seminal paper describing his experiments and his theory of genetics—"Versuche über Pflanzen-Hybriden" ("Experiments in Plant Hybridization")—to the Brünn (Moravia) Natural History Society in 1865. Is it possible that Darwin read that paper at some point? Were his letters to Wallace in 1866 inspired (to some extent at least) by Mendel's work rather than representing his own insights? If he had read Mendel's paper, why didn't he see that Mendel's results provided the definitive answer to Jenkin's criticism?

Intriguingly, no fewer than three books published between 1982 and 2000 alleged that copies of Mendel's paper had been found in Darwin's library, and a fourth book (published in 2000) even claimed that Darwin had supplied Mendel's name for inclusion in the *Encyclopaedia Britannica,* under an entry of "hybridism." Obviously, if this last claim were shown to be true, it would mean that Darwin was fully aware of Mendel's work.

Andrew Sclater of the Darwin Correspondence Project at Cambridge University responded definitively to all of these questions in 2003. As it turns out, Mendel's name (as an author) does not appear even once in the entire list of books and articles owned by Darwin. This is not surprising, given that Mendel's original paper appeared in the somewhat obscure proceedings of the Brünn Natural History Society, to which Darwin had never subscribed. Furthermore, Mendel's work languished almost unread for thirty-four years, until its rediscovery in 1900, when the botanists Carl Correns of Germany, Hugo de Vries of Holland, and Erich von Tschermak-Seysenegg of Austria published supporting evidence independently. Nevertheless, two of the books that were in Darwin's possession did refer to Mendel's work. In Darwin's book *The Effects of Cross and Self Fertilisation in the Vegetable Kingdom,* he even cited one of those books: Hermann Hoffmann's *Untersuchungen zur Bestimmung des Werthes von Species und Varietät* (*Examinations to Determine the Value of Species and Variety*), published in 1869. However, Darwin never cited Mendel's work, nor did he annotate any mention of Mendel in Hoffmann's book. Again, this is hardly surprising, since Hoffmann himself did not comprehend the significance of Mendel's work, and he summarized Mendel's conclusions by the rather low-key statement "Hybrids possess the tendency in the succeeding generations to revert to the parent species." Mendel's pea experiments were mentioned in another book owned by Darwin: *Die Pflanzen-Mischlinge* (*The Plant Hybrids*), by Wilhelm Olbers Focke. Figure 7 shows the title page, on which Darwin wrote his name. As I've seen with my own eyes, this book had an even less distinguished fate: The precise pages describing Mendel's work remained uncut in Darwin's copy of the book! (In old bookbinding, pages were connected at the outer edges and had to be cut open.) Figure 8 shows a picture of Darwin's copy, made at my request, displaying the uncut pages. However, had Darwin read those pages, he would not have been much enlightened, since Focke failed to grasp Mendel's principles.

One question still remains: Did Darwin indeed suggest Mendel's name to the *Encyclopaedia Britannica*? Sclater left no doubt as to the answer: No, he did not. Rather, when asked by the naturalist George Romanes to read a draft on hybridism for the *Britannica* and to provide references, Darwin sent him his copy of Focke's book (with the

Figure 7

uncut pages), suggesting to Romanes that the book could "aid you much better than I can!"

In contrast to Darwin's total lack of familiarity with Mendel's work, Darwin's theories did have a clear influence on Mendel's ideas, although not in 1854–55, when Mendel started his experiments with peas. Mendel possessed the second German edition of *The Origin*, which was published in 1863. In his copy, he marked certain passages by lines in the margin and others by underlining parts of the text. Mendel's markings show great interest in topics such as the sudden appearance of new varieties, artificial and natural selection, and differences between species. There is little doubt that reading *The Origin* had significantly affected Mendel's own writing by 1866, since

Figure 8

Mendel's paper reflects in many places various aspects of Darwin's concepts. For instance, in discussing the origin of heritable variation, Mendel wrote:

> If the change in the conditions of vegetation were the only cause for variability, one would expect that those cultivated plants which have been grown for centuries under almost constant conditions would revert again to stability. As is well known this is not the case, for it is precisely among such plants that not only the most varied, but also the most variable forms are found.

We can compare this language to that used in one of Darwin's paragraphs in *The Origin:* "No case is on record of a variable organism ceasing to vary under cultivation. Our oldest cultivated plants, such as wheat, still yield new varieties: our oldest domesticated animals are still capable of rapid improvement or modification." Most importantly, however, it seems that Mendel may have realized that his theory of heredity could solve Darwin's main problem: an adequate supply of heritable variations for evolution to influence. This was precisely where blending inheritance was failing, as pointed out by Jenkin. Mendel wrote:

> If it be accepted that the development of hybrids takes place in accordance with the Law established for *Pisum* [peas], each experiment must be undertaken with a great many individuals . . . In *Pisum* it has been proved by experiment that hybrids produce ovules and pollen-grains of differing constitutions, and that this is the reason for the variability of their offspring.

In other words, inherited variation, and no blending at all. Moreover, Mendel tried several times to create variations in plants by removing them from their natural habitat to his garden at the monastery. When this failed to produce any change, Mendel told his friend Gustav von Niessl, "This much already seems clear to me, that nature does not modify species in any such way, so some other force must be at work." Mendel therefore accepted at least some parts of the theory of evolution. This, however, raises another intriguing question: If Mendel agreed with Darwin's concepts, and maybe even recognized the importance of his own results for evolution, why didn't he mention Darwin by name in his writings? To answer this question we have to understand Mendel's special historical circumstances. On September 14, 1852, the Austrian Emperor Francis Joseph I empowered Prince-Bishop Rauscher to act as his representative in drawing up a concordat with the Vatican. This concordat was signed in 1855, and in reaction to the winds of change in Europe in 1848, it contained strict regulations such as: "All school instruction of Catholic children must be in accordance with the teachings of the Catholic Church . . . The bishops have the right to condemn books injurious to religion and morals, and to forbid Catholics reading them."

As a result of these restrictions, for instance, the paleontologist Antonén Frič was not even allowed to lecture in Prague, Czechoslovakia, on his impressions from a scientific meeting in Oxford in 1860 in which Huxley presented Darwin's theory. Although the Vatican itself delayed official pronouncements on Darwin's theory for many decades, a council of German Catholic bishops did state in 1860: "Our first parents were formed immediately by God, therefore we declare that the opinion of those who do not fear to assert that this human being . . . emerged finally from the spontaneous continuous change of imperfect nature to the more perfect, is clearly opposed to Sacred Scripture and to the Faith." In this rather oppressive atmosphere, Mendel, who was ordained a priest in 1847 and elected abbot of the monastery in 1868, probably did not think it prudent to express any explicit support for Darwin's ideas.

We may still wonder what might have happened had Darwin actually read Mendel's paper before November 21, 1866, when he completed his chapter on the ill-conceived pangenesis. Of course, we shall never know, but my guess is that nothing would have changed. Darwin was neither ready to think in terms of variation affecting just one part of an organism and not the others, nor was his mathematical ability sufficient to follow and fully appreciate Mendel's probabilistic approach. To develop a specific, universal mechanism from a few isolated cases of a 3:1 ratio in the transmission of some properties of a particular plant was not Darwin's forte. Moreover, Darwin's stubborn defense of his theory of pangenesis demonstrates that at that point in his life, he may have been afflicted with what modern psychologists refer to as the *illusion of confidence:* a common state in which people overestimate their abilities. While this principle generally applies to people who are unskilled but unaware of it, it can affect everybody at some level. For instance, studies show that most chess players think that they can play much better than their formal ranking implies. If Darwin indeed had the illusion of confidence, it would be quite ironic, since it was Darwin himself who once observed insightfully that "ignorance more frequently begets confidence than does knowledge."

The complexities of developing a quantitative approach to the phenomena of variation and rate of survival, and the complete integration of Darwinian selection and Mendelian genetics, took about seventy years to resolve. Initially, in the years following the 1900 rediscovery of Mendel's pioneering 1865 paper, Mendel's laws of heredity were even thought to be *opposed* to Darwinism. Geneticists argued that mutations—the only acceptable form of heritable variation—were abrupt and ready-made rather than gradually selectionist. This opposition subsided by the 1920s, following a number of seminal research projects. First, breeding experiments with the *Drosophila* fruit fly by biologist Thomas Hunt Morgan and his group demonstrated unambiguously that Mendel's principles were universal. Second, the geneticist William Ernest Castle was able to show that he could produce inherited change by the action of selection on small variations in traits in a population of rats. Finally, the English geneticist Cyril Dean Darlington discovered the actual mechanics of chromosomal exchange of genetic material. All of these, and similar studies, showed that mutations occurred infrequently and most of the time were disadvantageous. On those rare occasions in which advantageous mutations arose, natural selection was identified as the only mechanism that could enable their propagation through the population. Biologists further came to understand that a large number of separately acting genes could affect a continuous variation in a characteristic. Darwin's gradualism won the day, with natural selection acting on tiny differences to cause adaptation.

Darwin's blunder and Jenkin's criticism had another unexpected consequence: They essentially paved the way for the mathematical population genetics theory developed by Ronald Fisher, J. B. S. Haldane, and Sewall Wright. This was the work that provided the ultimate proof that Mendelian genetics and Darwinian selection were complementary and indispensable to each other. Given that Darwin got the fundamental fact of genetics wrong, it is absolutely amazing how much he got right.

The story of evolution is therefore not a simple narrative leading from myth to knowledge but a collection of diversions, blunders, and winding paths. Eventually, all of these intertwined threads converged into one conclusion: Understanding life requires understanding very intricate chemical processes that involve some very complex molecules. We shall pick up this important thread again in chapters 6 and 7, when we'll discuss the discovery of the molecular structure of proteins and of DNA.

I mentioned earlier the fact that Jenkin's article raised a few other objections to Darwin's theory of evolution. In particular, Jenkin relied on calculations by his friend and partner the famous physicist William Thomson (later Lord Kelvin), which appeared to show that the age of the Earth was much shorter than the vast expanses of time Darwin needed for his theory of evolution to work. The ensuing controversy provides us with fascinating insights not only into the differences between the methodologies in various branches of science but also (admittedly much more speculatively) into the operation of the human mind.

CHAPTER 4

HOW OLD IS THE EARTH?

In the Beginning, God created Heaven and Earth . . .
Which beginning of time, according to our chronology, fell
upon the entrance of the night preceding the twenty-third
day of October in the year of the Julian period, 710.

—JAMES USSHER, 1658

Humans have been curious about the age of the Earth since recorded history. It is not often the case, after all, that one number—the Earth's age—can have important implications for such diverse fields as theology, geology, biology, and astrophysics. Given that each one of these different disciplines has had its share of strongly opinionated individuals, we shouldn't be too surprised to discover that by the nineteenth century, the attempts to estimate the Earth's age had led to a number of bitter controversies.

The concept of a universal, linear time did not appear right away. In the ancient Hindu tradition, for instance, time had essentially no boundaries, and, like the ancient symbol of the ouroboros—the snake biting its own tail—the universe was assumed to undergo continuous cycles of destruction and regeneration. Nevertheless, the Hindu sages of antiquity did come up with a rather "precise" number for the Earth's age, which in 2013 was supposed to be 1,972,949,114 years. In the Western tradition, Plato and Aristotle were much more concerned with *why* and *how* the existing order of nature came about than with *when*, but even they toyed with the idea of recurring cycles, in step with the heavenly motions. In the Christian world, on the other hand, a circular time was rejected in favor of a unique, nonrepeating straight line, leading from creation to the Last Judgment. In this religious context, determinations of the age of the Earth had been for centuries the exclusive province of theologians. In one of the earliest estimates, Theophilus, the sixth bishop of Antioch, concluded in the year 169 that the world had been created some 5,698 years earlier. His motivation for calculating the age, he declared, was not "to furnish mere matter of much talk" but "to throw light upon the number of years from the foundation of the world." While Theophilus did allow for a certain margin of error in his calculation, he did not think the error would exceed 200 years.

Many of the chronologists that followed him tended to simply add up time intervals between key biblical events, the scriptural ages at death of certain individuals, or the span of generations. Prominent among these biblical scholars were John Lightfoot, the seventeenth-century vice chancellor of Cambridge University, and James Ussher, who became archbishop of Armagh in 1625. Even though the title of Lightfoot's 1642 short book was carefully constructed to read *A Few, and New Observations, upon the Book of Genesis: The Most of Them Certain, the Rest Probable, All Harmless, Strange, and Rarely Heard of Before,* Lightfoot did not hesitate to pronounce that the creation of the first human—Adam—occurred precisely at nine o'clock in the morning! As for the date of the creation of the world, Lightfoot settled on 3928 BCE.

Ussher's calculation was somewhat more sophisticated in that he complemented the biblical accounts by some astronomical and historical data. His punctilious conclusion: The world appeared on the evening before October 23, in the year 4004 BCE. This particular date became well known in the English-speaking world, since it was added as a marginal note to the English Bible in 1701.

Naturally, the Christian view of time followed largely upon the heels of the Jewish tradition, which was also based mostly on a literal reading of the narrative in the book of Genesis. In the context of a divine drama in which the Jewish people were supposed to play the principal role, having a history was clearly crucial. According to this heritage, the world was created some 5,773 years ago (as of 2013). Prophetically, one of the most influential Jewish scholars of the Middle Ages—Maimonides (Moshe ben Maimon)—advocated against a literal interpretation of the biblical text. As if anticipating what Galileo Galilei would say more than four centuries later, Maimonides argued that whenever accurate scientific findings are in conflict with the Scriptures, the biblical texts should be reinterpreted. The Dutch Jewish philosopher Baruch de Spinoza echoed the same sentiments: "The knowledge of . . . nearly everything contained in Scripture, must be sought only from Scripture itself, just as the knowledge of nature is sought from nature itself." In fact, Maimonides was not even the first to suggest that the passages in Genesis had been intended only allegorically. In the first century, the Hellenistic Jewish philosopher Philo Judaeus of Alexandria wrote presciently:

> It would be a mark of great naïvety to think that the world was created either in six days, or indeed in time at all; for time is nothing but the sequence of days and nights, and these things are necessarily connected with the motion of the Sun above and below the Earth. But the Sun is a part of the heavens, so that time must be recognized as something posterior to the world. So it would be correct to say not that the world was created in time, but that time owed its existence to the world.

As we shall see in chapter 10, Philo's last sentence conforms nicely to Einstein's ideas in his theory of general relativity.

The great German philosopher Immanuel Kant was one of the first to judge critically the balance between the biblical interpretation and the laws of physical science. Kant himself leaned decisively toward physics. He pointed out in 1754 the danger of relying on the human lifetime in estimating the age of the Earth. Kant wrote, "Man makes the greatest mistake when he tries to use the sequence of human generations that have passed in a particular [period of] time

as a measure for the age of the greatness of God's works." Referring
to a sarcastic passage written by the French author Bernard le Bovier
de Fontenelle in 1686, in which roses were metaphorically ponder-
ing the age of their gardener, Kant added a "citation" from the roses:
"Our gardener, is a very old man; in rose memory he is just the same
as he has always been; he doesn't die or even change."

Around the same time that Kant was ruminating on the nature of
existence, the French diplomat and geologist Benoît de Maillet carried out one of the first bold attempts to use actual observations and
methodical scientific reasoning to determine the age of the Earth. De
Maillet took advantage of his position as French general consul at
various spots around the Mediterranean to make geological observations which convinced him that the Earth could not have been created fully formed in one instant of time. Rather, he inferred a long
history of gradual geological processes. Being fully aware of the
risks involved in challenging the dominance of the church's ortho-
doxy, de Maillet composed his theory on the history of the Earth
in a series of manuscripts that were collected, edited, and published
under the title of *Telliamed* ("de Maillet" in reverse) only in 1748,
ten years after de Maillet's death. The work was written as a fictional
string of conversations between an Indian philosopher (named
Telliamed) and a French missionary. While de Maillet's original ideas
have been somewhat watered down by the tinkering of his editor,
the Abbott Jean Baptiste le Mascrier, it is still possible to discern the
basic argument. In modern terms, this was a theory of what is now
known as sedimentation. Fossilized shells in sedimentary rocks near
mountaintops led de Maillet to conclude that water entirely covered
the young Earth. This hypothesis offered a potential solution to a
question Leonardo da Vinci had already agonized over two centuries earlier: "Why the bones of great fishes and oysters and corals
and various other shells and sea-snail are found on the high tops of
mountains that border on the sea, in the same way in which they
are found in the depths of the sea?" De Maillet married his idea of a
water-covered Earth with René Descartes's theory of the solar system—in which the Sun resided in a vortex about which the planets

were swirling—to say that the Earth was losing its water into the vortex. Having observed in several ancient ports such as Acre, Alexandria, and Carthage a rate of decline of the sea level by about three inches per century, de Maillet was able to estimate an age for the Earth of about 2.4 billion years.

Strictly speaking, de Maillet's calculations and the theory on which they were based were flawed in a number of ways. First, water never entirely covered the Earth—de Maillet did not realize that rather than the water receding, the land might rise. Second, his understanding of rock formation was seriously lacking. He further weakened his case by occasional wanderings into fantasy. For instance, to support his contention that all life-forms emerged from the sea (an idea that is actually consistent with present thinking), de Maillet relied on accounts of mermaids and men with tails. Nevertheless, de Maillet's estimate of the age of the Earth marked a major shift in the thinking about this problem. For the first time, it was not the human lifetime by which the age of the Earth was determined but rather the rate of natural processes.

De Maillet humbly dedicated his book to the romantic French dramatist Cyrano de Bergerac, who died less than a year before de Maillet's birth. He started his dedication this way: "I hope you will not take it ill, that I address my present work to you, since, I could not possibly have made choice of a more worthy Protector of the Romantic Flights of Fancy which it contains." Today we can appreciate that de Maillet's work was more than "romantic flights of fancy"—it contained the seeds of geochronology. Determining the age of the Earth by scientific methods was about to become a worthy intellectual challenge.

The Earth and Life Gain a History

In his masterwork *Principia*, first published in 1687, Isaac Newton noted that "a globe of red hot iron equal to our earth, that is, about 40,000,000 feet in diameter, would scarcely cool in an equal number of days, or in above 50,000 years." Realizing he could not easily square this result with his religious beliefs, Newton was quick to add, "But I suspect that the duration of heat may, on account of some latent causes, increase in a yet less proportion than that of the diameter; and I should be glad that the true proportion was investigated by experiments."

Newton was not the only seventeenth-century scientist to think about this problem. The famous philosophers Descartes and Gottfried Wilhelm Leibniz also discussed the cooling of the Earth from an initially molten state. However, the first person who appears to have taken seriously Newton's advice about an experimental investigation—and who in addition was imaginative enough to attempt to use the cooling problem to estimate the age of the Earth—was the eighteenth-century mathematician and naturalist Georges-Louis Leclerc, Comte de Buffon.

Buffon was a truly prolific character who was not only an accomplished scientist but also a successful businessman. He is perhaps best known for the clarity and forcefulness with which he presented a new method for approaching nature. His monumental lifework, *Histoire Naturelle, Générale et Particulière* (*Natural History, General and Particular*)—thirty-six-volumes of which were completed during his lifetime (with eight more published posthumously)—was read by most of the educated people of the day in Europe and North America. Buffon's aim was to deal in succession with topics ranging from the solar system, the Earth, and the human race to the different kingdoms of living creatures.

In his mental excursion into the Earth's physical past, Buffon assumed that the Earth started as a molten sphere after having been ejected from the Sun due to a collision with a comet. Then, in the true spirit of an experimentalist, he was not satisfied with a purely theoretical scenario—Buffon proceeded immediately to manufacture spheres of different diameters and to measure accurately the time it took them to cool down. From these experiments he estimated that the terrestrial globe solidified in 2,905 years and cooled down to its present temperature in 74,832 years, even though he suspected that the cooling time could be much longer.

Eventually, however, it was not pure Newtonian physics that brought the problem of the Earth's age into the limelight. The surge in the study of fossils in the eighteenth century convinced naturalists such as Georges Cuvier, Jean-Baptiste Lamarck, and James Hutton that both the paleontological and the geological records required the operation of geological forces over exceedingly long periods of time. So long, in fact, that, as Hutton has put it, he found "no vestige of a beginning, no prospect of an end."

In view of the increasing difficulty of trying to cram the entire history of the Earth into the biblical mere few thousand years, some of the more religiously inclined naturalists (but not only them) opted to rely on catastrophes such as floods as agents of rapid changes. If great expanses of time were to be denied, catastrophes appeared to be the only vehicle that could significantly shape the Earth's surface almost instantaneously. To be sure, the distribution of marine fossils provided clear evidence for the action of flooding and glaciation in the Earth's geological past, but many of the ardent catastrophists were at least partially motivated by their unwavering loyalty to the biblical text rather than by the scientific attestation. Richard Kirwan—one of the well-known chemists of the day—articulated this position clearly. Kirwan pitted Hutton directly against Moses in describing how dismayed he was to observe "how fatal the suspicion of the high antiquity of the globe has been to the credit of Mosaic history, and consequently to religion and morality."

The situation started to change dramatically with the publication of Charles Lyell's three-volume *Principles of Geology* in the years 1830–33. Lyell, who was also Charles Darwin's close friend, made it clear that the catastrophist doctrine was far too frail to last as a compromise between science and theology. He decided to put aside the question of the origin of the Earth and to concentrate on its evolution. Lyell argued that the forces that sculpted the Earth— volcanism, sedimentation, erosion, and similar processes—remained essentially unchanged throughout the Earth's history, both in their strength and in their nature. This was the idea of uniformitarianism that inspired Darwin's concept of gradualism in the evolution of species. The basic premise was simple: If there was one thing that these slow-acting geological forces needed in order to have an appreciable effect, it was time. Lots of it. Lyell's followers have almost abandoned the notion of a definite age altogether in favor of the rather vague "inconceivably vast" time. In other words, Lyell's Earth was one that was almost in a *steady state,* with snail's-pace changes operating over a nearly infinite time. This principle starkly contrasted with the theological estimates of some six thousand years.

To a certain extent, the world view of an immeasurably extended geological age permeated Darwin's *The Origin,* even though Darwin's own attempt to estimate the age of the Weald—the eroded valley stretching across the southeastern part of England—turned out to be disastrously flawed, and he eventually retracted it. Darwin envisaged for evolution a long sequence of phases, lasting perhaps ten million years each. There was, however, one important difference between Darwin's stance and those of the geologists. While he indeed required long periods of time for evolution to run its course, he insisted on a directional "arrow of time"; he could not be satisfied with a steady state or a cyclical progression, since the concept of evolution gave time a clear trend. But a controversy was starting to brew. It was not between Darwin and Lyell personally, nor even between geology and biology in general, but between a champion of physics on one side and some geologists and biologists on the other. Enter one of the most eminent physicists of his time: William Thomson, later known as Lord Kelvin.

Global Cooling

In 1897 the *Vanity Fair Album,* a compendium of highlights from the weekly British society magazine, published a eulogy of Lord Kelvin, part of which read as follows:

> His father was Professor of Mathematics at Glasgow. Himself was born in Belfast seventy-two years ago, and educated at Glasgow University and at St Peter's, Cambridge;

of which College, after making himself Second Wrangler and Smith's Prizeman, he was made a Fellow. Unlike a Scotchman, he presently returned to Glasgow—a Professor of Natural Philosophy; and since then he has invented so much and, despite his mathematical knowledge, has done so much good, that his name—which is William Thomson—is known not only throughout the civilized world but also on every sea. For when he was a mere knight he invented Sir William Thomson's mariner's compass as well as a navigational sounding machine, that is, unhappily less well known. He has also done much electrical service at sea: as engineer for various Atlantic cables, as inventor of the mirror-galvanometer and siphon recorder, and much else that is not only scientific but useful. He is so good a man, indeed, that four years ago he was enobled as Baron Kelvin of Largs; yet he is still full of wisdom, for his Peerage has not spoiled him . . . He knows all there is to know about heat, all that is yet known about Magnetism, and all that he can find out about Electricity. He is a very great, honest, and humble Scientist who has written much and done more.

This was a fairly accurate, if humorous, description of the numerous accomplishments of the man dubbed by one of his biographers the "Dynamic Victorian." On his ennoblement, in 1892, Thomson adopted the title Baron Kelvin of Largs, after the River Kelvin, which flowed close to his laboratory at the University of Glasgow. "Second Wrangler" referred to Kelvin having placed (to his disappointment) second in the final honors school of mathematics at Cambridge. Story has it that on the morning the examination results were to be posted, he sent his servant to find out "who is Second Wrangler?" and was devastated when he was told "You, sir!" There is no doubt that Kelvin was the foremost figure of the age that witnessed the end of classical physics and the birth of the modern era. Figure 9 shows a portrait of Lord Kelvin, possibly after a photograph taken in 1876.

Figure 9

Appropriately, upon his death in 1907, he was laid to rest in a tomb alongside Isaac Newton in Westminster Abbey. What the eulogy did not capture, however, was the eventual collapse of Kelvin's stature in scientific circles. As an old man, Kelvin developed a reputation as an obstructionist to modern physics. Often portrayed as someone who clung stubbornly to his old views, he resisted new findings about atoms and about radioactivity. More surprisingly, even though James Clerk Maxwell relied on some of Kelvin's applications of energy principles when he developed his impressive theory of electromagnetism, Kelvin still objected to the theory, stating, "I may say that the one thing about it that seems intelligible to me, I do not think is admissible." For the technically savvy person that he was, Kelvin made similarly astonishing declarations on technology, such as "I have not the smallest molecule of faith in aereal navigation other than ballooning." It was this enigmatic man—brilliant as a young scientist, seemingly out of touch as an old one—who attempted to discredit the geologists' views on the age of the Earth.

On April 28, 1862, Kelvin (then still Thomson) read to the Royal Society of Edinburgh a paper entitled "On the Secular Cooling of the Earth." This paper followed closely on the heels of another article published just the month before, with the title "On the Age of the Sun's Heat." Thomson made clear from the opening sentence that this was not going to be just another forgettable technical essay. Here was a hard-line attack on the geologists' assumption about the unchanging nature of the forces that had shaped the Earth:

> For eighteen years it has pressed on my mind, that essential principles of Thermodynamics have been overlooked by those geologists who uncompromisingly oppose all paroxysmal hypotheses, and maintain not only that we have examples now before us, on the earth, of all the different actions by which its crust has been modified in geological history, but that these actions have never, or have not on the whole, been more violent in past time than they are at present.

While the phrase "pressed on my mind" was somewhat of an overdramatized exaggeration, it was certainly true that Kelvin's first papers on the topics of heat conduction and the distribution of heat through the body of the Earth were written as early as 1844 (when he was a twenty-year-old student) and 1846, respectively. Even before his seventeenth birthday, Thomson succeeded in spotting a mistake in a paper on heat by an Edinburgh professor.

Kelvin's point was simple: Measurements from mines and wells indicated that heat was flowing from the Earth's interior to its surface, implying that the Earth was an initially hotter planet that was cooling. Consequently, Kelvin argued, unless some internal or external energy sources could be shown to compensate for the heat losses, clearly no steady state, or repeating, identical geological cycles, were possible. Charles Lyell was actually aware of this problem, and in his *Principles of Geology* he proposed a self-sustaining mechanism by which he believed that chemical, electric, and heat energy could be exchanged cyclically in the Earth's interior. Basically, Lyell envisaged a scenario in which chemical reactions generated heat, which drove electrical currents, which in turn dissociated the chemical compounds into their original constituents, thus starting the process anew. Kelvin could barely hide his contempt. He demonstrated unambiguously that such a process amounted to some sort of perpetual motion machine, violating the principle of dissipation (and conservation) of energy—when mechanical energy is transformed irreversibly into heat, as in the case of friction. Lyell's mechanism therefore violated the basic laws of thermodynamics. To Kelvin, this was the ultimate proof that the geologists were completely ignorant of physical principles, and he remarked caustically:

> To suppose, as Lyell, adopting the chemical hypothesis, has done, that the substances, combining together, may be again separated electrolytically by thermoelectric currents, due to the heat generated by their combination, and thus the chemical action and its heat continued in an endless cycle, violates the principles of natural philosophy in exactly the same manner, and to the same degree, as to believe that a clock constructed with a self-winding movement may fulfill the expectations of its ingenious inventor by going for ever.

At its core, Kelvin's calculation of the age of the Earth was straightforward. Since the Earth was cooling, he explained, one could use the science of thermodynamics to calculate the Earth's finite geological age: the time it took the Earth to get to its current state, since the formation of the solid crust. The idea itself was not entirely new; the French physicist Joseph Fourier had developed the mathematical theory of thermal conductivity and of the Earth's cooling process at the beginning of the nineteenth century. Realizing the theory's potential, Kelvin engaged in 1849 in a series of measurements of underground temperatures (together with the physicist James David Forbes), and in 1855 urged that a complete geothermal survey be conducted, precisely to enable the calculation of the Earth's age.

Kelvin assumed that the mechanism that transported heat from the interior to the surface was the same type of conduction that transfers heat from an iron skillet on an open fire to its handle. Still, in order to apply Fourier's theory to the cooling Earth, he needed to know three physical quantities: (1) the initial internal temperature of the Earth, (2) the rate of change in the temperature according to depth, and (3) the value of the thermal conductivity of the Earth's rocky crust (which determines how fast heat can be transported).

Kelvin thought that he had a fairly good handle on two of these quantities. Measurements by a number of geologists have shown that while results varied from location to location, in the mean, the temperature toward the Earth's center increased roughly by one degree Fahrenheit for every fifty feet of descent (this quantity is known as the temperature gradient). Concerning the thermal conductivity, Kelvin relied on his own measurements for two types of rocks and for sand to give him what he regarded as an acceptable average. The third physical quantity—the Earth's deep internal temperature—was extremely problematic, since it couldn't be measured directly. But Kelvin was not a man easily deterred by such difficulties. Putting his analytic mind to work, he was eventually able to deduce an estimate for the unknown internal temperature. The entangled intellectual maneuvering that he had to perform to achieve this result presented Kelvin at his best—and his worst. On one hand, his virtuosic command of physics and his ability to examine potential alternatives with a razor-sharp logic were second to none. On the other, as we shall see in the next chapter, due to his overconfidence, he could sometimes be completely blindsided by unforeseen possibilities.

Kelvin started his assault on the problem of the Earth's internal temperature by analyzing a variety of possible models for the cooling Earth. The general assumption was that the Earth's initial state was molten, as a result of the heat generated by some collision—either with a number of smaller bodies, such as meteors, or with one body of nearly equal mass. The subsequent evolution of this molten sphere depended on a property of rocks that was not known with certainty: whether upon solidifying, molten rock expanded (as in the case of freezing water) or contracted (as metals do). In the former case, one could expect the solid crust to float over a liquid interior, just like ice on the surface of lakes in winter. In the latter, the denser solid rocks forming near the Earth's cooler surface would have sunk down, eventually forming perhaps a solid scaffolding that could support the surface crust. While the empirical evidence was scarce, experiments with melted granite, slate, and trachyte all seemed to point in the direction of molten rock contracting both upon cooling and solidifying. Kelvin used this information to chart a new scenario. He proposed that before complete solidification took place, the cooler surface liquid had sunk toward the center, thus maintaining convection currents similar to those generated in the oil in a frying pan. In this model the convection was assumed to sustain a nearly uniform temperature throughout. Consequently, Kelvin assumed that at the point of solidification, the temperature everywhere was roughly the temperature at which rock melts, and he took that to be the Earth's internal temperature (assuming that the core had not cooled by much since). This model implied that the Earth was nearly homogeneous in its physical properties. Unfortunately, even this ingenious scheme did not fully solve the problem, since the value of the fusion temperature of rock was not known in Kelvin's time. He was, therefore, forced to adopt an educated guess of seven thousand to ten thousand degrees Fahrenheit for an acceptable range. (Seismic measurements performed in 2007 gave a temperature of about 6,700 degrees Fahrenheit for a region that is about 1,860 miles below the Earth's surface.)

Putting together all of this information, Kelvin finally computed an age for the Earth's crust: ninety-eight million years. Estimating the uncertainties in his assumptions and in the data available to him, Kelvin believed that he could state with some confidence that the Earth's age had to be somewhere between twenty million and four hundred million years.

In many respects, in spite of the insecure assumptions, this was a truly brilliant calculation. Who would have thought that one could actually calculate the age of the Earth? Kelvin took a seemingly insoluble problem and deciphered it. He used sound physical principles both in the formulation of the problem and in his method of calculation, and he augmented those by the best quantitative measurements available at the time (some of which he performed by himself). Compared with his determination, the geologists' estimates appeared to be nothing more than crude guesses and idle speculation based on poorly understood processes such as erosion and sedimentation.

The number that Kelvin produced—roughly one hundred million years—was broadly consistent with an earlier estimate he had made of the age of the Sun. This was significant, since even some of Kelvin's contemporaries realized that the strength of his argument about the age of the Earth derived at least part of its credibility from his solar calculation. Kelvin's basic premise in the paper "On the Age of the Sun's Heat," and in a few similar later papers, was not very different from his central thesis in his analysis of the age of the Earth. The key assumption was that the *only* source of energy that the Sun had at its disposal was the mechanical *gravitational energy.* This was supposed to be supplied either by the falling-in of meteors, as Kelvin originally thought, but later rejected, or, as Kelvin proposed later and forcefully reiterated in 1887, by the Sun continually contracting, and dissipating its gravitational energy in the form of heat. Since, however, the energy supply was clearly not infinite, and the Sun was unceasingly losing energy by radiation, Kelvin concluded justifiably that the Sun could not remain unchanged indefinitely. To calculate its age, he borrowed elements from theories for the formation of the solar system proposed by the French physicist Pierre-Simon Laplace and the German philosopher Immanuel Kant. He then supplemented those with important insights on the Sun's potential contraction gained from the work of his contemporary German physicist Hermann von Helmholtz. Weaving all of these ingredients into one coherent picture, Kelvin was able to obtain a rough estimate of the Sun's age. The last paragraph in Kelvin's paper reflected his acknowledgement of the many uncertainties involved:

> It seems, therefore, on the whole most probable that the sun has not illuminated the earth for one hundred million years, and almost certain that he has not done so for five

hundred million years. As for the future, we may say, with equal certainty, that inhabitants of the earth can not continue to enjoy the light and heat essential to their life for many million years longer unless sources now unknown to us are prepared in the great storehouse of creation.

As I shall describe in the next chapter (and explain in detail in chapter 8), the last sentence proved to be truly farseeing.

The fact that the calculated ages of the Sun and the Earth turned out to be comparable—even though the estimates were determined independently—made Kelvin's calculation more compelling, since there was every reason to suspect that the entire solar system had formed around the same time. Still, quite a few British geologists remained unconvinced. It almost seemed as though, for some of them, it was more convenient to explain everything not by the laws of physics but rather by what the American geologist Thomas Chamberlin cynically termed in 1899 "reckless drafts on the bank of time." The best illustration of the skeptical attitude toward Kelvin's findings is a fascinating exchange Kelvin had in 1867 with the Scottish geologist Andrew Ramsay. The occasion was a lecture by the geologist Archibald Geikie on the geological history of Scotland. Kelvin later described the conversation he had with Ramsay immediately following the talk, noting that almost every word of it remained "stamped on my mind":

> I asked Ramsay how long a time he allowed for that history. He answered that he could suggest no limit to it. I said, "You don't suppose geological history has run through 1,000,000,000 [one billion] years?" "Certainly I do!" "10,000,000,000 [ten billion] years?" "Yes!" "The sun is a finite body. You can tell how many tons it is. Do you think it has been shining for a million million years?" "I am as incapable of estimating and understanding the reasons which you physicists have for limiting geological time as you are incapable of understanding the geological reasons

for our unlimited estimates." I answered, "You can understand the physicists' reasoning perfectly if you give your
mind to it."

Kelvin was absolutely right. Ignoring for a moment the question
of how solid his physical assumptions were and the mathematical
details of his calculations, Kelvin's main point was accessible. Since
the Sun and the Earth are both losing energy, and they don't possess
any known sources that could replenish the losses, he argued, the
Earth's geological past must have been more active than the present.
A hotter Sun would have caused more evaporation, with the associated higher rate of erosion by precipitation. At the same time, a
hotter Earth would have experienced heightened volcanic activity.
Consequently, Kelvin concluded, the uniformitarian assumption of
an Earth in an almost indefinite quasi–steady state was untenable.

It wasn't surprising, then, that in 1868, when Kelvin delivered an
address before the Geological Society of Glasgow, he chose as the target for his acrimonious criticism the first text that had brought the
principle of uniformitarianism (formulated by James Hutton) to the
attention of a wide audience. This was the 1802 book *Illustrations
of the Huttonian Theory of the Earth* by the Scottish scientist John
Playfair. From this book, Kelvin cited the following stunning passage,
which to him represented the epitome of the orthodox opinion of the
geologists of the day:

> How often these vicissitudes of decay and renovation have
> been repeated it is not for us to determine; they consti-
> tute a series of which, as the author of this theory [Hut-
> ton] has remarked, we neither see the beginning nor the
> end, a circumstance that accords well with what is known
> concerning other parts of the economy of the world . . .
> in the planetary motions where geometry has carried the
> eye so far both into the future and the past, we discover
> no mark either of the commencement or the termination
> of the present order. *It is unreasonable, indeed, to suppose*

that such marks should exist anywhere [emphasis added]. The Author of nature has not given laws to the universe which, like the institutions of men, carry in themselves the elements of their own destruction. He has not permitted in His works any symptoms of infancy or of old age, or any sign by which we may estimate either their future or their past duration. He may put an end, as He, no doubt, gave a beginning to the present system, at some determinate time; but we may safely conclude that the great *catastrophe* will not be brought about by any of the laws now existing, and that it is not indicated by anything which we perceive.

Kelvin's reaction to this excerpt was merciless. "Nothing," he said, "could possibly be further from the truth." Explaining again his argument in layman's terms, he noted:

The earth, if we bore into it anywhere, is warm, and if we could apply the test deep enough, we should, no doubt, find it very warm. Suppose you should have here before you a globe of sandstone, and boring into it found it warm, boring into another place found it warm, and so on, would it be reasonable to say that that globe of sandstone has been just as it is for a thousand days? You should say, 'No; that sandstone has been in the fire, and heated not many hours ago.' It would be just as reasonable to take a hot water jar, such as is used in carriages, and say that that bottle has been as it is for ever — as it was for Playfair to assert that the earth could have been for ever as it is now, and that it shows no traces of a beginning, no progress towards an end.

To strengthen his argument further, Kelvin decided not to rely just on his old reasoning about the Earth and the Sun. He came up with yet a third line of evidence, based on the Earth's rotation around its axis. The concept itself was ingenious and easy to understand. An initially molten Earth would have assumed, due to its spin, a slightly oblate shape: more flattened at the poles and more bulging at the equator. The faster the initial spin, the less spherical the resulting shape. This form, Kelvin inferred, would have been preserved upon the Earth's solidification. Precise measurements of the deviation from sphericity could therefore be used to determine the original rate of rotation. Since tides caused by the Moon's gravity were expected to act like friction and slow down the rotation, one could estimate how much time was required to reduce the initial spin rate to the present one rotation every twenty-four hours.

While the idea was fascinating, turning it into an actual number for the age of the Earth was extremely tricky. Kelvin himself admitted, "It is impossible, with the imperfect data we possess as to the tides, to calculate how much their effect in diminishing the earth's rotation really is." Nevertheless, Kelvin felt that even the mere fact that one *could* place a limit on the age of the Earth, no matter how uncertain, was sufficient to refute the uniformitarian notion of an inconceivably vast time. Referring to his own numerical estimate of a 22-second retardation per century in the Earth's rotation period, he concluded, "[Whether] the earth's lost time is 22 seconds, or considerably more or less than 22 seconds, in a century, the principle is the same. There cannot be uniformity. The earth is filled with evidences that it has not been going on for ever in the present state, and that there is a progress of events toward a state infinitely different from the present."

Disappointingly for Kelvin, the estimate based on the Earth's spin rate did not last for very long, at least not in any quantitative way. As fate would have it, none other than George Howard Darwin, Charles Darwin's fifth child, showed the argument to be useless for an age estimate. George was a physicist with considerable mathematical dexterity. He attacked the problem of the spinning Earth with infinite patience and attention to detail. In a series of papers published mainly between 1877 and 1879, the younger Darwin was able to demonstrate that, contrary to Kelvin's expectation, the Earth could continue to gradually change its shape even as its rotation rate was slowing down. This was a consequence of the fact that even a solidified Earth was not completely rigid. The bottom line was unequivocal. Darwin showed that given the many uncertainties about the Earth's interior, there was no reliable way to calculate the planet's age from its spin.

Needless to say, Charles Darwin was delighted to discover that his own son managed to "stagger" the great Kelvin, and he exclaimed, "Hurrah for the bowels of the earth and their viscosity and for the moon and for the Heavenly bodies and for my son George."

But George Darwin's work did not affect Kelvin's main claims — it only established that Kelvin's third argument (concerning the Earth's rotation) could not be used to support the *value* of the estimate of the age of the Earth. There was another sense, however, in which Darwin's work was revealing. It showed that even the august Lord Kelvin was not infallible. As we shall see in the next chapter, this may have helped to open the door for further criticism.

Deep Impact

To describe the age-of-the-Earth controversy as a battle to the death between physics and geology would be a mistake. While certainly there was tension along disciplinary lines, Kelvin saw himself so much in the mainstream of British geology that in his address at the Glasgow Geological Society meeting in 1878 he did not hesitate to declare, "*We, the geologists* [emphasis added], are at fault for not having demanded of the physicists experiments on the properties of matter." This "flexible" self-identification reflected the less compartmentalized scientific world of the nineteenth century. Victorian scientists freely attended meetings of societies that formally represented other branches of science. Rather than a dispute between disciplines, therefore, the age-of-the-Earth debate was largely a clash between Kelvin and the doctrine of *some* geologists.

One may wonder what it was that motivated Kelvin to examine this problem in the first place. The answer is actually quite simple. Even a cursory examination leaves little doubt that the publication of Darwin's *The Origin* in 1859 provided the main impetus for Kelvin's direct attack on the estimates of the ages of both the Sun and the Earth. To be clear, Kelvin did not object to the theory of evolution per se. In his 1871 presidential address to the British Association for the Advancement of Science, for instance, he expressed moderate support for some of Darwin's conclusions in *The Origin*. However, he completely rejected natural selection because he had "always felt that this hypothesis does not contain the true theory of evolution, if evolution there has been, in biology." Why not? Because, he explained, he was "profoundly convinced that the argument of design has been greatly too much lost sight of in recent zoological speculations." In other words, even this committed mathematical physicist, who passionately declared that the "essence of science . . . consists in inferring antecedent conditions, and anticipating future evolutions, from phenomena which have actually come under observation," still believed that "overpoweringly strong proofs of intelligent and benevolent design lie all around us." In fact, Kelvin held that the laws of thermodynamics themselves were a part of that universal design. Nevertheless, we should remember that even if Kelvin felt a certain emotional attachment to the concept of "design," there is no doubt that he anchored his scathing criticism of the geologists' practices in genuine physics, not in his religious beliefs.

What was Kelvin's impact on geology? Until the 1860s, geologists were much more preoccupied with discussions on whether the Earth's interior was solid or fluid than with the Earth's chronology. By the mid-1860s, however, quite a few of the influential geologists started to pay serious attention to Kelvin's claims. Foremost among these were John Phillips, Archibald Geikie, and James Croll. Based on studies of sediments, Phillips himself had suggested in 1860 an age of about ninety-six million years for the Earth. By 1865, he was publicly supporting Kelvin. Geikie, the new director of the Geological Survey for Scotland, more or less assumed the role of a conduit and mediator between physics and geology. On one hand, he criticized Kelvin's assertion that the Earth's geological past had been more active, citing evidence that seemed to show that if anything "the intensity . . . has, on the whole, been augmenting." On the other, in a paper published in 1871, he essentially abandoned uniformitarianism and stated that based on research in physics "about 100 millions of years is the time assigned within which all geological history must be comprised." Croll, an impressive self-taught physicist and geologist, was entirely convinced by Kelvin's calculation of the cooling Earth, and even though he was extremely skeptical about Kelvin's estimate of the age of the Sun, accepted one hundred million years for the Earth's age.

Often you can judge whether a certain scientific theory has had an impact by the vehemence with which the heavyweights with something at stake announce their objections to it. In Kelvin's case, the sure sign that the opposition had taken notice came when biologist Thomas Huxley attacked Kelvin's calculation in February 1869.

Huxley had earned the title "Darwin's Bulldog" because of his aggressive support of the theory of evolution and his eagerness to debate in its defense. Huxley loved controversy as much as Darwin hated it. He is perhaps best known for his legendary, brief verbal encounter with Samuel Wilberforce, bishop of Oxford, on June 30, 1860. The event took place at Oxford University's New Museum library as part of the annual conference of the British Association for the Advancement of Science. The story was told in colorful, although probably partly imaginary, detail in the October 1898 issue of *Macmillan's Magazine*. The writer reminisced:

> I was happy enough to be present on the memorable occasion at Oxford when Mr. Huxley bearded Bishop Wilberforce . . . Then the Bishop rose, and in a light scoffing tone, florid and fluent, he assured us there was nothing in the idea of evolution; rock-pigeons were what rock-pigeons had always been. Then, turning to his antagonist with a smiling insolence, he begged to know, was it through his grandfather or his grandmother that he claimed his descent from a monkey? On this Mr. Huxley slowly and deliberately arose. A slight tall figure stern and pale, very quiet and very grave, he stood before us, and spoke those tremendous words—words which no one seems sure of now,

nor I think, could remember just after they were spoken, for their meaning took away our breath, though it left us in no doubt as to what it was. He was not ashamed to have a monkey for his ancestor; but he would be ashamed to be connected with a man who used great gifts to obscure the truth. No one doubted his meaning and the effect was tremendous. One lady fainted and had to be carried out.

Even though there are many versions of the precise wording of this impromptu exchange, Huxley's oratorical skills and the rising sentiments against the interference of men of the church with science have helped this legend grow. Historian of science James Moore even went so far as to state, "No battle of the nineteenth century, since Waterloo, is better known."

Huxley decided to come to the geologists' defense in his 1869 presidential address to the Geological Society of London. First, he took advantage of the fact that Kelvin's condemnation happened to be directed at the rather old text by Playfair to make the questionable claim "I do not suppose that, at the present day any geologists would be found to maintain absolute Uniformitarianism." He continued by asking rhetorically whether any geologist had ever required more than one hundred million years for the action of geology. This was really a sleight of hand, since Huxley's own "master," Darwin himself, wrongly estimated an age of three hundred million years for the Weald. Finally, after a few more dubious, if eloquent, assertions, Huxley pronounced his own summary that "the case against [geology and biology] has entirely broken down."

Huxley's address drew a furious response from one of Kelvin's most staunch supporters: Peter Guthrie Tait. The mathematician, who never missed an opportunity for a good fight, wrote a review of Kelvin's and Huxley's addresses in which he wrapped insults directed at Huxley in a few polite sentences. Then, to deliver an even more punishing blow, Tait decided to cite a number for the age of the Earth that not only had no physical justification whatsoever but also was even shorter than Kelvin's most extreme estimates:

> We find that we may, with considerable probability, say
> that Natural Philosophy already points to a period of some
> ten or fifteen millions of years as all that can be allowed for
> the purpose of the geologist and paleontologist; and that
> it is not unlikely that, with better experimental data, this
> period may be still further reduced.

The net result of Tait's provocative statements was an increasing sense of discontent among the geologists, who felt that in spite of their efforts to come to terms with Kelvin's limitations, the physicists were not reciprocating by any concessions to geological evidence. These details notwithstanding, however, there was no question that, conceptually at least, Kelvin had won the battle, and the notion of a limited rather than immeasurable time for the age of the Earth had triumphed. By the end of the nineteenth century, the idea of a steady state Earth had given way to the realization that calculating the age of the Earth using physical principles was part of what geology was all about.

You might have thought that these enormous gains to geology, combined with Kelvin's other myriad contributions to science (he published more than six hundred papers), would have exalted him to the rank of those who have had an everlasting impact—the likes of Galileo and Newton. Sadly, the reality is rather different, and even the fact that Kelvin was equally comfortable in the academic and technical worlds did not help. In 1999 *Physics World* magazine and *Physics Web* (an internet publication by the British Institute of Physics) conducted polls in which they asked one hundred leading physicists to name the ten greatest physicists of all time. Kelvin's name did not feature on either list. At least one of the reasons for this subsequent deterioration in Kelvin's status concerns the debate over the age of the Earth: We know today that the age of the Earth is about 4.54 billion years. *This is about fifty times longer than Kelvin's estimate!* How could he have blundered so badly in a calculation supposedly based on the laws of physics?

CHAPTER 5

CERTAINTY GENERALLY IS ILLUSION

Science becomes dangerous only when it imagines that it
has reached its goal.

— GEORGE BERNARD SHAW

The debate on the age of the Earth between Kelvin and
Thomas Huxley generated considerable scientific and public
interest. Few disagreed with the assessment that, if anything,
Kelvin's position had been strengthened somewhat by this war of
words. Huxley did raise one point, however, which proved to be
particularly perceptive. In effect, it identified the crux of Kelvin's
blunder:

> Mathematics may be compared to a mill of exquisite work-
> manship, which grinds you stuff of any degree of fineness;
> but, nevertheless, what you get out depends upon what
> you put in; and as the grandest mill in the world will not
> extract wheat-flour from peascod, so pages of formulae
> will not get a definite result out of loose data.

Indeed, Kelvin had such an exceptional command of mathematics that it was essentially guaranteed that if he had made any mistake,

it would not have been in the actual calculations. It was his set of *assumptions* that provided the input for those calculations that had to be scrutinized.

The Gutsy Pupil

The first person who, albeit reluctantly, took a stab at searching for a loophole in Kelvin's original postulates was Kelvin's former pupil and assistant, the engineer John Perry. By happenstance, Perry studied engineering under James Thomson, Kelvin's older brother, but later he spent a year in Kelvin's Glasgow laboratory. While most of Perry's scientific output focused on electrical engineering and applied physics, he is perhaps best known today for his brief foray into geology.

In August 1894, Robert Cecil, the Third Marquis of Salisbury, delivered a presidential address at the sixty-fourth meeting of the British Association for the Advancement of Science. Salisbury used Kelvin's estimate of the age of the Earth (one hundred million years) to argue that evolution by natural selection could not have taken place. As is often the case with messages that are too dogmatic, however, this speech achieved precisely the opposite effect to its intentions, at least with John Perry. Salisbury's denial of the theory of evolution convinced Perry that there had to be something wrong with Kelvin's calculations. Impressed by the accumulation of geological and paleontological data, Perry wrote to a physicist friend that "once it became clear to my mind that there was necessarily such a flaw [in Kelvin's estimates], its discovery was no mere question of chance."

Perry completed the first version of his investigation of the problem of the cooling Earth on October 12, and during the following weeks, he diligently sent copies of the paper to a number of physicists (including Kelvin) for comments. Respectful even in his criticism, Perry signed his letter to Kelvin "Your affectionate pupil." While about a half dozen physicists expressed support for Perry's conclusions, Kelvin himself did not bother to respond. Perry was given a second chance when he was invited to a dinner party at Trinity College in Cambridge, a dinner that Kelvin was also supposed to attend. The opportunity to talk to Kelvin in person was too good to be missed. Perry excitedly described the event to a friend the following day:

> I sat beside him [Kelvin] last night at Trinity and he had to listen. I knew beforehand that he would not read my documents and he hadn't but I gave him a lot to think of and his pitying smile at my ignorance died away in about 15 minutes. I think he will now really begin to consider the matter. Geikie [the geologist Archibald Geikie] was opposite, his eyes gleaming with delight.

The scientific journal *Nature* eventually published Perry's article on January 3, 1895. The report started in an apologetic tone: "I have sometimes been asked by friends interested in geology to criticize Lord Kelvin's calculation of the probable age of the earth. I have usually said that it is hopeless to expect that Lord Kelvin should have made an error in the calculation." Perry then went on to express his personal reservations about the methodology used in geology at the time: "I dislike very much to consider any quantitative problem set by a geologist. In nearly every case the conditions given are much too vague for the matter to be in any sense satisfactory, and a geologist does not seem to mind a few millions of years in matters relating to time." Finally, Perry explained what had nevertheless convinced him to take on the daunting task of challenging Kelvin: "His [Kelvin's] calculation is just now being used to discredit the direct evidence of geologists and biologists, and it is on this account that I have considered it my duty to question Lord Kelvin's conditions."

Perry focused most of his attention on one of Kelvin's basic assumptions: that the Earth's conductivity was *the same* at all depths. In other words, Kelvin assumed that heat was transported with uniform efficiency, be it at a depth of one mile or a thousand miles. This hypothesis was crucial. Just as a forensic investigator can determine the time of death by measuring the temperature of the corpse, Kelvin used this assumption to determine the cooling age of the Earth, by measuring by how much the temperature within the Earth increased with each foot of depth. Kelvin's calculation showed that if the Earth were older than about one hundred million years, then the temperature would rise with depth more slowly than was actually observed, since the cooled skin would be thicker.

Perry wondered, what if instead of being the same everywhere, heat transport in the deep interior were more effective than near the surface? Clearly, in that case, the bottom of the Earth's outer skin could be kept warmer for much longer. In particular, Perry showed that if the Earth's interior happened to be partly fluid, then, just as with water heated in a deep pot, heat could be *convected* to the surface crust so efficiently (by the fluid itself) that the age estimate could be extended even to three billion years. He then concluded his article by addressing the arguments based on the age of the Sun and on the Earth's spin, but there was nothing really new in his discussion of these topics. Regarding the question of the tidal retardation of the Earth's rotation rate, Perry called attention primarily to George Darwin's demonstration that even a solid Earth could still alter its shape.

At first, Perry's article (in its prepublication circulated form) drew a response not from Kelvin himself but from his self-appointed "bulldog": Peter Guthrie Tait. In an offensively dismissive letter, Tait wrote to Perry on November 22, 1894:

> ... my entire failure to catch the *object* of your paper. For I seem to gather that you don't object to Lord Kelvin's mathematics. Why, then, drag in mathematics at all, since it is absolutely obvious that the better conductor the interior in comparison with the skin, the longer ago must it have been when the whole was at 7000°F [Kelvin's assumed melting point for rocks]: the state of the skin being as at present? I don't suppose Lord Kelvin would care to be troubled with a demonstration of *that*.

Tait appears to have missed the point entirely. Since no one at the time could tell with any certainty what the conditions at great depths really were, what one *assumed* for the purpose of the calculation was a matter of mere conjecture. Perry's intention was simply to show that if one made a different assumption from Kelvin's about the Earth's interior—that at the bowels of the Earth heat was transported more easily than in the Earth's outer skin—then calculations based on physical principles could be made compatible with the old age that the geologists and biologists were requiring. Kelvin's blunder was in *not realizing that the latitude allowed by the existing observations could introduce a much larger uncertainty into his estimated age than he was willing to acknowledge.*

In his response to Tait, Perry attempted to be polite, noting, "You say I am right, and you ask my object. Surely Lord Kelvin's case is lost, as soon as one shows that there are *possible* [emphasis added] conditions as to the internal state of the earth which will give many times the age which is your and his limit." In language probably reflecting the admiration of a former assistant, he then added, "What troubles me is that I cannot see one bit that you have reason on your side, and yet I have been so accustomed to look up to you and Lord Kelvin, that I think I must be more or less of an idiot to doubt when you and he were so 'cocksure.'"

The conciliatory tone was apparently lost on Tait, since he continued to retort disparagingly: "I should like to have your answers to *two* questions: (1) What grounds have you for supposing the inner materials of the earth to be better conductors than the skin?" The second "question" was not really a question but a contemptuous remark about the insatiable expectations of the geologists: "(2) Do you fancy that any of the *advanced* geologists would thank you for the ten billion years instead of one hundred million? Their least demand is for a trillion:—for *part* of the mere secondary period!" (Figure 10 shows a copy of his note.) But Perry did not give up: "It is for Lord Kelvin to prove that there is not greater conductivity inside," he insisted.

It goes without saying that Perry was correct in his assessment.

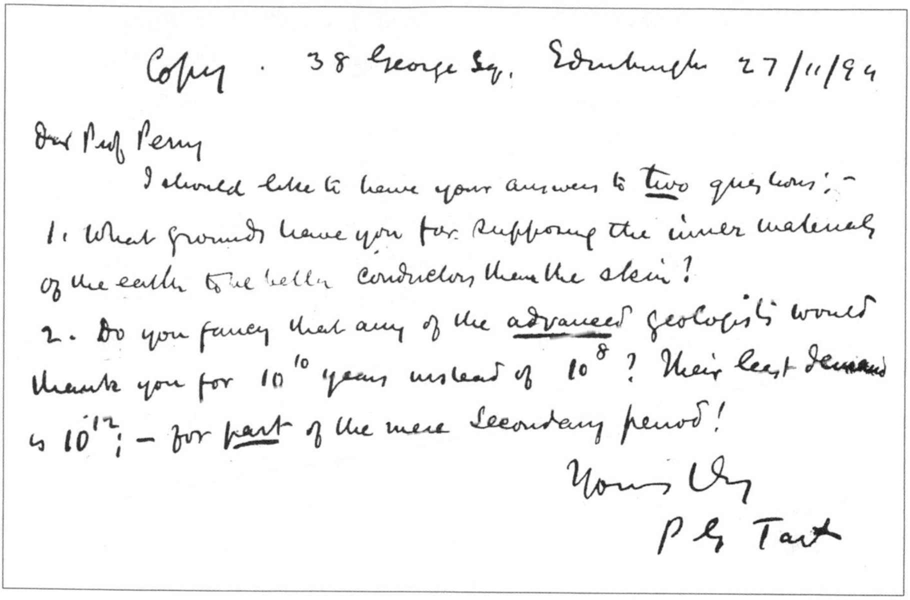

Figure 10

In the absence of any definitive experimental evidence as to the Earth's precise internal conditions, the fact that he was able to show that Kelvin *could* be wrong by a large factor was sufficient.

When he finally decided to respond, Kelvin was much less aggressive than Tait. While he stated, "I feel that we cannot assume as in any way probable the enormous differences of conductivity and thermal capacity at different depths which you [Perry] take for your calculations," he also remarked in an uncharacteristically appeasing style, "I thought my range from 20 millions to 400 millions was probably wide enough, but it is quite possible that I should have put the superior limit a good deal higher, perhaps 4000 instead of 400." Perhaps at no other time did Kelvin show such respect for opinions that contradicted his own. Most likely this magnanimity expressed his sense of obligation to empathize with a former student. He hastened to insist, however, that his estimate of the Sun's age was still "refusing sunlight for more than a score or a very few scores of million years of past time." As we shall see later in this chapter, Kelvin had no reason at the time to revise his calculation for the age of the Sun.

Perry's challenge caused Kelvin to spend the following couple of months conducting experiments with heated basalt, marble, rock salt, and quartz. These experiments seemed to show, in agreement with new results by the Swiss geologist Robert Weber, that the conductivity either did not change much or even decreased slightly with increasing temperature. Unfortunately for Perry, Weber's new results contradicted those of his own previous experiments—the very experiments Perry had used to support his case. The overjoyed Kelvin published the results in *Nature* on March 7, 1895, breaking the news that "Prof. Perry and I had not to wait long . . . to learn that there was no ground for the assumption of greater conductivity of rock at higher temperatures." Kelvin further cited a conclusion of the American geologist Clarence King, who (without considering the possibility of convection by a fluid) stated: "We have no warrant for extending the earth's age beyond 24 millions of years." Kelvin pronounced gleefully that he was "not led to differ much from his [King's] estimate of 24 million years."

Perry, however, was not convinced. Concentrating on *possible* internal conditions, rather than trying, like Kelvin, to guess what the most *probable* conditions might be, he noted that King's conclusion was still constrained by the assumption of a solid, homogeneous Earth. In a paper that appeared in *Nature* on April 18, 1895, Perry summarized his views on the impasse: "Now it is evident that if we take any probable law of temperature of convective equilibrium at the beginning and assume that there may be greater conductivity inside than on the surface rocks, Mr. King's ingenious test for liquidity will not bar us from almost any great age." Perry's logic was clear: His goal was to demonstrate that the Earth could be older than Kelvin's estimate, even if he was unable to identify the precise flaw in Kelvin's argument, due to uncertainties concerning the Earth's internal structure. The measurements of the conductivity of heated rocks might have disproved one of the ways in which heat could be transported more readily at great depths, but other possibilities were still open. In particular, convection by fluidlike mass was an attractive alternative.

Perry's intuition turned out to be visionary. He continued to maintain that the failure of Kelvin's model to produce greater ages was a direct consequence of Kelvin's assumption of a homogeneously conductive Earth, and that this limitation could be overcome if one allowed the Earth's mantle to convect. It took the geologists of the twentieth century a few decades to prove Perry right. The realization that convection was possible, even within what appeared to be a rather solid mantle, played an important role in the eventual acceptance of the idea (first introduced in 1912 by the German scientist Alfred Wegener) of plate tectonics and continental drift. Not only can heat be transported by fluidlike motion but also entire continents can move horizontally over long periods of time. The precise conditions at the interface between the Earth's inner core and the outer part continues to be a hot topic (no pun intended) of research even today.

Perry concluded his last article on the subject of the age of the Earth with an unambiguous statement:

> From the three physical arguments [tidal retardation of the Earth's spin; the cooling of the Earth; and the age of the Sun], Lord Kelvin's higher limits are 1,000, 400, and 500 million years. I have shown that we have reasons for believing that the age, from all three, may be very considerably under estimated. It is to be observed that if we exclude everything but the arguments from mere physics, the *probable* age of life on the earth is much less than any of the above estimates; but if the palaeontologists have good reasons for demanding much greater times, I see nothing from the physicist's point of view which denies them four times the greatest of these estimates.

Perry saw nothing wrong with 4 billion years for the Earth's age, fairly close to today's determination of about 4.5 billion years.

Perry's work created the first crack in Kelvin's seemingly unshakable calculations, by challenging the postulates that Kelvin made

concerning the Earth's solidity and homogeneity. There was, however, another crucial hypothesis in Kelvin's estimate of the age of
the Earth: that there were no unknown internal or external energy
sources that could compensate for the heat losses. Events toward the
end of the nineteenth century demolished this premise too.

Radioactivity

In the spring of 1896, the French physicist Henri Becquerele discovered that the decay of unstable atomic nuclei is accompanied by
spontaneous emission of particles and radiation. The phenomenon
became known as *radioactivity*. Seven years later, physicists Pierre
Curie and Albert Laborde communicated that the decay of radium
salts provided a previously unknown source of heat. It took the amateur astronomer William E. Wilson less than four months from the
Curie and Laborde announcement to come up with the speculation that this property of radium "may possibly afford a clue to the
source of energy in the sun and stars." Wilson estimated that just "3.6
grams of radium per cubic meter of the sun's volume would supply
the entire output." While Wilson's extremely short note to *Nature*
received relatively little attention from the scientific community, the
potential implications of an unanticipated source of energy did not
escape George Darwin. This mathematical physicist, who ceaselessly
looked for ways to free geology from the straitjacket imposed by
Kelvin's chronology, declared emphatically in September 1903: "The
amount of energy available [in radioactive materials] is so great as
to render it impossible to say how long the sun's heat has already
existed, or how long it will last in the future." The Irish physicist and
geologist John Joly embraced this pronouncement enthusiastically
and immediately applied it to the problem of the age of the Earth. In
a letter to *Nature* published on October 1, Joly pointed out that "a
source of supply of heat [the radioactive minerals] in every element
of material" would be equivalent to an increased transfer of heat
from the Earth's interior. This was precisely what Perry had shown
was needed in order to increase the age estimates. Put differently, in

Kelvin's scenario, the Earth was merely losing heat from its original reservoir. The discovery of a new source of internal heat seemed to undermine the entire basis for this scheme.

One of the key figures in the ensuing frantic research on radioactivity was the young New Zealand–born physicist Ernest Rutherford, who later became known as the "father of nuclear physics." At the time, Rutherford was working at McGill University in Montreal (he later moved to the United Kingdom), where he concluded on the basis of scores of experiments that the atoms of all of the radioactive elements contained enormous amounts of latent energy that could be released as heat. One journal welcomed the announcement by Rutherford that the Earth would survive much longer than Kelvin had estimated with the headline: "DOOMSDAY POSTPONED."

On his part, Kelvin showed great interest in the discoveries concerning radium and radioactivity, but he remained unconvinced that these would alter his age estimates. Refusing to admit, at least initially, that the source of energy of the radioactive elements could come from within, he wrote, "I venture to suggest that somehow ethereal waves may supply the energy to the radium while it is giving out heat to the ponderable matter around it." In other words, Kelvin proposed that the atoms simply collect energy from the ether (ether was supposed to permeate all space), only to release it back upon their decay. In 1904, however, with considerable intellectual courage, he abandoned this idea at the British Association meeting, although he never published a retraction in print. Unfortunately, for some unclear reason, he again lost touch with the rest of the physics community in 1906 when he rejected the notion that radioactive decay transmuted one element into another, even though Rutherford and others had accumulated solid experimental evidence for this phenomenon. Throughout this period, Rutherford's one-time collaborator Frederick Soddy lost his patience. In an acerbic exchange with Kelvin in the pages of the London *Times,* he declared disrespectfully, "It would be a pity if the public were misled into supposing that those who have not worked with radioactive bodies [alluding to Kelvin] are as entitled to as weighty an opinion as those who have." Even before that altercation, in a book he had published in 1904, Soddy did not hesitate to firmly assert that "the limitations with respect to the past and future history of the universe have been enormously extended."

Rutherford was a little more generous. Many years later, he told and retold an anecdote related to a lecture on radioactivity that he had given in 1904 at the Royal Institution:

> I came into the room, which was half dark, and presently spotted Lord Kelvin in the audience and realized that I was in for trouble at the last part of the speech dealing with the age of the Earth, where my views conflicted with his. To my relief he fell fast asleep but as I came to the important point, I saw the old bird sit up, open an eye and cock a baleful glance at me! Then sudden inspiration came, and I said Lord Kelvin had limited the age of the Earth, *provided no new source of heat was discovered.* That prophetic utterance refers to what we are now considering tonight, radium! Behold! The old boy beamed at me.

Eventually, *radiometric dating* became one of the most reliable techniques to determine the ages of minerals, rocks, and other geological features, including the Earth itself. Generally, a radioactive element decays into another radioactive element at a rate determined by its *half-life:* the period of time it takes for the initial amount of radioactive material to decrease by half. The decay series continues until it reaches a stable element. By measuring and comparing the relative abundances of naturally occurring radioactive isotopes and all of their decay products, and coupling those data with the known half-lives, geologists have been able to determine the Earth's age to high precision. Rutherford was one of the pioneers of this technique, as the following story documents: Rutherford was walking on campus with a small black rock in his hand, when he met his Canadian geologist colleague Frank Dawson Adams. "Adams," he asked, "how old is the earth supposed to be?" Adams answered that several methods had given an estimate of one hundred million years. Rutherford then commented quietly, "I know that this piece of pitchblende [a mineral that is a source of uranium] is seven hundred million years old."

Most if not all descriptions of the age-of-the-Earth controversy would have you believe that Kelvin's dramatically wrong age estimate was a direct consequence of the fact that he ignored radioactivity. If this were the whole truth, Kelvin's error would not have qualified as a blunder in my book, since Kelvin could not have considered a previously undiscovered source of energy. However, it is actually mistaken to attribute the erroneous age determination entirely to radioactivity. It is true that radioactive decays within the entire volume of the Earth's mantle (down to a depth of about 1,800 miles) do indeed produce heat at a rate that is roughly equal to half the rate of heat flow through the planet. But not all of this heat can be tapped readily. A careful examination of the problem reveals that, given Kelvin's assumptions, had he even included radioactive heating, he really should have considered only the heat generated inside the Earth's outer 60-mile-deep skin. The reason is that Kelvin showed that only heat from such depths could be effectively mined by *conduction* in about one hundred million years. Geologists Philip England, Peter Molnar, and Frank Richter demonstrated in 2007 that when this fact is taken into account, the inclusion of radioactive heat deposition would not have altered Kelvin's estimate for the age of the Earth in any significant way. Kelvin's most serious blunder was not in being unaware of radioactivity (even though, once discovered, ignoring it was certainly not justified), but in initially ignoring and later objecting to the possibility raised by Perry of convection within the Earth's mantle. This was the true source of the unacceptably low age estimate.

How could a man of such intellectual powers as Kelvin be so sure that he was right even when he was dead wrong? Like all humans, Kelvin still had to use the hardware between his ears—his brain—and the brain has limitations, even when it belongs to a genius.

On the Feeling of Knowing

Since we can neither interview Kelvin nor image areas of his functioning brain, we will never know for sure the precise reasons for his misguided stubbornness. We do know, of course, that people who have spent much of their working lives defending certain propositions do not like to admit that they were wrong. But shouldn't have Kelvin, the great scientist that he was, been different? Isn't changing one's theories based on new experimental evidence part of what science is all about? Fortunately, modern psychology and neuroscience are beginning to shed some light on what has been termed the "feeling of knowing," which almost certainly shaped some of Kelvin's thinking.

I should first note that in his approach to science and crusade for knowledge, Kelvin was more akin to an engineer than to a philosopher. Being on one hand an effective mathematical physicist, and on the other, a gifted experimentalist, he always sought a premise with which he could calculate or measure something rather than an opportunity to contemplate different possibilities. At the very basic level, therefore, Kelvin's blunder was a consequence of his belief that he could always determine what was *probable*, not realizing the ever-present danger of overlooking some possibilities.

At a somewhat deeper stratum, Kelvin's blunder probably stemmed from a well-recognized psychological trait: The more committed we are to a certain opinion, the less likely we are to relinquish it, even if confronted with massive contradictory evidence. (Does the phrase "weapons of mass destruction" ring a bell?) The theory of *cognitive dissonance*, originally developed by psychologist Leon Festinger, deals precisely with those feelings of discomfort that people experience when presented with information that is inconsistent with their beliefs. Multiple studies show that to relieve cognitive dissonance, in many cases, instead of acknowledging an error in judgment, people tend to reformulate their views in a new way that justifies their old opinions.

The messianic stream within the Jewish Hasidic movement

known as Chabad provided an excellent, if esoteric, example for this process of reorientation. The belief that the Chabad leader Rabbi Menachem Mendel Schneerson was the Jewish Messiah picked up momentum during the decade preceding the rabbi's death in 1994. After the Rabbi suffered a stroke in 1992, many faithful followers in the Chabad movement were convinced that he would not die but would "emerge" as the Messiah. Faced with the shock of his eventual death, however, dozens of these followers changed their attitudes and argued (even during the funeral) that his death was, in fact, a *required* part of the process of his returning as the Messiah.

An experiment conducted in 1955 by psychologist Jack Brehm, then at the University of Minnesota, demonstrated a different manifestation of cognitive dissonance. In that study, 225 female sophomore students (the classical subjects of experiments in psychology) were first asked to rank eight manufactured articles as to their desirability on a scale of 1.0 ("not at all desirable") to 8.0 ("extremely desirable"). In the second stage, the students were allowed to choose as a take-home gift one of two articles presented to then from the original eight. A second round of rating all eight items then followed. The study showed that in the second round, the students tended to increase their ratings for the article they had chosen and to lower them for the rejected item. These and other similar findings support that idea that our minds attempt to reduce the dissonance between the cognition "I chose item number three" and the cognition "But item number seven also has some attractive features." Put differently, things seem better after we choose them; a conclusion corroborated further by neuroimaging studies that show enhanced activity in the caudate nucleus, a region of the brain implicated with "feeling good."

Kelvin's case appears to fit the cognitive dissonance theory like a glove. After having repeated the arguments about the age of the Earth for more than three decades, Kelvin was not likely to change his opinion just because someone suggested the *possibility* of convection. Note that Perry was not able to prove that convection was taking place, nor was he even able to show that convection was probable. By the time radioactivity appeared on the scene another decade later, Kelvin was probably even less inclined to publish a concession of defeat. Instead, he preferred to engage in an elaborate scheme of experiments and explanations intended to demonstrate that his old estimates still held true.

Why is it so difficult to let go of opinions, even in the face of contradictory evidence that any independent observer would regard as convincing? The answer can perhaps be found in the way the reward circuitry of the brain operates. Already in the 1950s, researchers James Olds and Peter Milner of McGill University identified pleasure centers in the brains of rats. Rats were found to press the lever that activated the electrodes placed at these pleasure-inducing locations more than six thousand times per hour! The potency of this pleasure-producing stimulation was illustrated dramatically in the mid-1960s, when experiments showed that when forced to choose between obtaining food and water or the rewarding pleasure stimulation, rats suffered self-imposed starvation.

Neuroscientists of the past two decades have developed sophisticated imaging techniques that allow them to see in detail which parts of the human brain light up in response to pleasing tastes, music, sex, or winning at gambling. The most commonly used techniques are positron-emission tomography (PET) scans, in which radioactive tracers are injected and then followed in the brain, and functional MRI (fMRI), which monitors the flow of blood to active neurons. Studies showed that an important part of the reward circuitry is a collection of nerve cells that originate near the base of the brain (in an area known as the ventral tegmental area, or VTA) and communicate to the nucleus accumbens—an area beneath the frontal cortex. The VTA neurons communicate with the nucleus accumbens neurons by dispatching a particular chemical neurotransmitter called dopamine. Other brain areas provide the emotional contents and relate the experience to memories, and to the triggering of responses. The hippocampus, for instance, effectively "takes notes," while the amygdala "grades" the pleasure involved.

So how does all of this relate to intellectual endeavors? To embark on, and persist in, some relatively long-term thought process, the brain needs at least some promise of pleasure along the way. Whether it is the Nobel Prize, the envy of neighbors, a salary raise, or the mere satisfaction of completing a Sudoku puzzle labeled "evil," the nucleus accumbens of our brain needs some dose of reward to keep going. However, if the brain derives frequent rewards over an extended period of time, then just as in the case of those self-starving rats, or with people who are addicted to drugs, the neural pathways connecting the mental activity to the feeling of accomplishment become gradually adapted. In the case of drug addicts, they need more drugs to get the same effect. For intellectual activities, this may result in an enhanced need for being right all the time and, concomitantly, in an increasing difficulty to admit errors. Neuroscientist and author Robert Burton suggested specifically that the insistence upon being right might have physiological similarities to other addictions. If true, then Kelvin would no doubt match the profile of an addict to the sensation of being certain. Almost a half century of what he surely regarded as victorious battles with the geologists would have strengthened his convictions to the point where those neural links could not be dissolved. Irrespective, however, of whether the sensation of being certain is addictive or not, fMRI studies have shown that what is known as motivated reasoning—when the brain converges on judgments that maximize positive affect states associated with attaining motives—is not associated with neural activity linked to cold reasoning tasks. In other words, motivated reasoning is regulated by emotions, not by dispassionate analysis, and its goal is to minimize threats to the self. It is not inconceivable that late in life, Kelvin's "emotional mind" occasionally swamped his "rational mind."

You may recall that earlier I referred to Kelvin's calculation of the age of the Sun. I do not consider his estimate to be a blunder. How is that possible? After all, his estimate of less than one hundred million years was wrong by as much as his value for the age of the Earth.

Fusion

In an article on the age of the Earth written in 1893, three years before the discovery of radioactivity, the American geologist Clarence King wrote, "The concordance of results between the ages of the sun and earth certainly strengthens the physical case and throws the burden of proof upon those who hold to the vaguely vast age derived from sedimentary geology." King's point was well taken. As long as the age of the Sun was estimated to be only a few tens of millions of years, any age estimates based on sedimentation would have been constrained, since for sedimentation to occur, the Earth had to be warmed by the Sun.

Recall that Kelvin's calculation of the age of the Sun relied entirely on the release of gravitational energy in the form of heat as the Sun contracts. This idea—that gravitational energy could be the source of the Sun's power—originated with the Scottish physicist John James Waterston as early as 1845. Ignored initially, the hypothesis was revived by Hermann von Helmholtz in 1854, and then enthusiastically endorsed and popularized by Kelvin. With the discovery of radioactivity, many assumed that the radioactive release of heat would turn out to be the real source of the Sun's power. This, however, proved to be incorrect. Even under the wild assumption that the Sun is composed largely of uranium and its radioactive decay products, the power generated would not have matched the observed solar luminosity (as long as chain reactions not known at Kelvin's time were not included). Kelvin's estimate of the age of the Sun had served to strengthen his objection to revising his calculation of the age of the Earth—as long as the problem of the age of the Sun existed, the discrepancy with the geological guesstimates would not have been resolved fully. The answer to the question of the Sun's age came only a few decades later. In August 1920, astrophysicist Arthur Eddington suggested that the *fusion* of hydrogen nuclei to form helium might provide the energy source of the Sun. Building on this concept, physicists Hans Bethe and Carl Friedrich von Weizsäcker analyzed a variety of nuclear reactions to explore the viability of this hypothesis. Finally, in the 1940s, astrophysicist Fred Hoyle (whose groundbreaking work we shall investigate in chapter 8) proposed that fusion reactions in stellar cores could synthesize the nuclei between carbon and iron. As I noted in the previous chapter, Kelvin was therefore right when he declared in 1862: "As for the future, we may say, with equal certainty, that inhabitants of the earth can not continue to enjoy the light and heat [of the Sun] essential to their life for many million years longer *unless sources now unknown to us are prepared in the great storehouse of creation [emphasis added]*." The solution to the problem of the age of the Sun required no less than the combined genius of Einstein, who showed that mass could be converted into energy, and the leading astrophysicists of the twentieth century, who identified the nuclear fusion reactions that could lead to such a conversion.

In spite of the fact that Kelvin's calculation of the age of the Earth was a blunder, I continue to regard it as absolutely brilliant. Kelvin had completely transformed geochronology from vague speculation into an actual science, based on the laws of physics. His pioneering work opened a vital dialogue between geologists and physicists— an exchange that continued until the discrepancy was resolved. At the same time, Kelvin's parallel work on the age of the Sun pointed clearly to the need to identify new sources of energy.

Charles Darwin himself was very aware of the importance of eliminating the obstacle to his theory presented by Kelvin's calculations. In his final revision of *The Origin*, Darwin wrote:

> With respect to the lapse of time not having been sufficient since our planet was consolidated for the assumed amount of organic change, and this objection, as urged by Sir William Thomson [Kelvin], is probably one of the gravest as yet advanced, I can only say, firstly, that we do not know at what rate species change as measured by years, and secondly, that many philosophers are not as yet willing

to admit that we know enough of the constitution of the universe and of the interior of our globe to speculate with safety on its past duration.

Darwin did not live to see how Perry's idea of a convective Earth, the discovery of radioactivity, and the understanding of nuclear fusion reactions in stellar interiors swept away all of Kelvin's age limits. The fact remains, however, that it was Kelvin's calculation—erroneous though it was—that identified the problem that had to be solved.

From our perspective as humans, one of the key benefits of the Earth having enjoyed 4.5 billion years of energy from the Sun has been the emergence of complex life on Earth. But the building blocks of all life-forms are cells, and by the 1880s, scientists using ever-improving optics to examine the internal structure of cells coined the term "chromosome" for the stringy bodies found in the cell's nucleus. Soon thereafter, Mendel's work on genes ("factors," as he called them) was rediscovered, and pioneering work by Thomas Hunt Morgan and his students at Columbia University allowed for mapping out the positions of genes along chromosomes. In 1944 a particular molecule—DNA—located on chromosomes, started to gain attention. Before long, biologists realized that all cells receive their instructions not from proteins but from two molecules, DNA and RNA nucleic acids. Biologists identified the DNA molecules as the bosses of all the frenzied activity in cells and the molecules that know how to make identical copies of themselves. RNA (ribonucleic acid) molecules were shown to be in charge of transmitting the instructions issued by DNA molecules to the rest of the cell. Together, these molecules contain all the information needed to make an apple tree, a snake, a woman, or a man function. The discoveries of the molecular structures of proteins and of DNA are two of the most fascinating stories in the search for the origin and workings of life. Yet these discoveries, too, involved major blunders.

CHAPTER 6

INTERPRETER
OF LIFE

In the fields of observation, chance favors only the mind
that is prepared.

—LOUIS PASTEUR

The lecture hall in the Kerckhoff Laboratory building at Caltech had rarely been as packed as it was that day in December 1950. Rumor had it that the famous chemist Linus Pauling was about to reveal something truly dramatic—maybe even a solution to one of life's greatest mysteries. When Pauling finally arrived, one of his research assistants was carrying an object that looked like a large sculpture, covered by a piece of cloth fastened with a string. The lecture itself demonstrated yet again Pauling's virtuosic command of chemistry, coupled with his exquisite showmanship. After keeping his audience in suspense for a while, Pauling finally used his jackknife to cut the string and, like a magician producing a rabbit out of a hat, unveiled what has become known as the *alpha-helix*: a three-dimensional stick-and-ball model of the main structural feature of many proteins.

One of the people who soon after heard about Pauling's pyrotechnical talk, even though at the time he was thousands of miles away in Geneva, Switzerland, was James Watson, who only three years later would discover (with Francis Crick) the structure of DNA. Watson was visiting the Swiss molecular biologist Jean

Weigle, who happened to be just back from spending a winter at Caltech. Even though Weigle could not quite judge the correctness of Pauling's multicolored wooden model, his report on the dazzling lecture was sufficient to intrigue and embolden Watson. We shall return to that gripping story later in the chapter.

By September 1951, the account of Pauling's scientific achievement had made it even into the pages of *Life* magazine, where a photograph of a grinning Pauling pointing to his alpha-helix model was accompanied by the headline "Chemists Solve a Great Mystery: Protein Structure Is Determined." The *Life* article was but a brief summary, in lay terms, of what had been a truly miraculous year in Pauling's long career. Suffice it to note that the May 1951 issue of the *Proceedings of the National Academy of Sciences* contained no fewer than seven papers by Pauling and his collaborator, chemist Robert Corey, on the topic of the structure of proteins ranging from collagen (the most abundant protein in mammals) to the shafts of feathers. This publication marked the culmination of fifteen years of trailblazing research by Pauling.

The Road to the Alpha-Helix

Pauling started to think about proteins in the 1930s. His first papers on the subject proposed a theory for hemoglobin—the iron-containing protein in red blood cells—suggesting that each of the four iron atoms in the molecule formed a chemical bond with an oxygen molecule. While working on that subject, Pauling pioneered a new experimental technique. He came up with the idea that measuring the magnetic properties of some proteins could provide important information on the nature of the bonds formed by iron atoms with the groups surrounding them. The method has indeed proved to be a fruitful tool in structural chemistry. Pauling used the magnetic characteristics to good effect; for instance, to determine the rates of several chemical reactions.

Around the same time, Alfred Mirsky, a leading protein expert, came to Pasadena for a year to work with Pauling's group. This chance collaboration between the two scientists became the starting point for an immensely successful quest. Mirsky and Pauling first proposed that a native protein—that is, an unaltered protein in its natural state inside the cell—is composed of chains of amino acids known as *polypeptides,* which are folded in some regular fashion. Very soon thereafter, Pauling realized that a key question was the precise nature of this folding. Fortunately, a few clues were starting to emerge in the early 1930s from X-ray diffraction experiments. In this powerful technique, scientists shine an X-ray beam onto a crystal. Then they can attempt to reconstruct the structure of the crystal (in terms of distances between atoms and their mutual orientations) from the way the invisible rays bounce off the sample. Pauling had at his disposal X-ray diffraction patterns obtained by the physicist William Astbury from hair, wool, horns, and fingernails (proteins known as *alpha keratin*). The X-ray photographs were rather fuzzy, however, and they did not allow for reliable structure determinations. Nevertheless, the photos did appear to indicate that the structural unit was repeating along the hair's axis every 5.1 angstroms. (One angstrom is a unit of length equal to one hundred-millionth of a centimeter.) Given the relatively poor quality of the X-ray patterns, Pauling decided to attack the problem from the other end: to use structural chemistry—the expected interactions among atoms— to predict the dimensions and shape of the polypeptide chain, and then to check which one of the various potential configurations was consistent with the information deduced from the X-ray images.

Pauling immersed himself in the work on the folding riddle in the early summer of 1937, when he was finally free of his teaching duties. Figure 11 shows a schematic drawing of the type of general structure that he was considering. By scrutinizing carefully the chemical bond between the carbon atom (denoted by "C" in the figure) and its adjacent nitrogen atom (denoted by "N"), Pauling concluded that the carbon, nitrogen, and the four neighboring atoms (collectively known as the peptide group) had to lie *in the same plane*. This particular feature turned out to be extremely important because it restricted greatly the number of possible structures, and

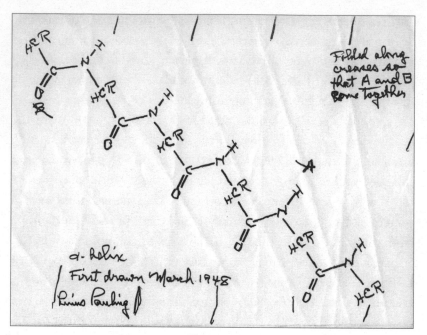

Figure 11

Pauling therefore hoped to be able to pin down the correct configuration. Science, however, rarely proceeds precisely as expected. In spite of several weeks of very intensive work, Pauling was unable to find a way of folding the peptide chains that would reproduce the repeat every 5.1 angstroms along the fiber axis that the X-ray results seemed to indicate. Frustrated, he gave up at that point.

When a promising hypothesis doesn't quite work, scientists often attempt to improve the quality of the available experimental data, since superior information can reveal previously indiscernible pointers. In this spirit, Pauling convinced Robert Corey to embark on a long-term project intended to determine the structure of some simple peptides and amino acids—the building blocks of proteins—using X-ray crystallography. Corey plunged into this study wholeheartedly, and by 1948, he and his collaborators at Caltech managed to unearth the exact architecture of about a dozen such compounds. Realizing that all of Corey's findings about the chemical bond lengths and the angles between different parts of the molecules, as

well as the *planarity* (the atoms lying in the same plane) of the peptide group, agreed precisely with his own previous formulations,
Pauling decided to revisit the problem of the structure of the alpha
keratin protein. In an account dictated on his (by then ancient) Dic-
taphone in 1982, Pauling recalled the circumstances:

> In the spring of 1948, I was in Oxford, England, serving as
> George Eastman Professor for the year and as a fellow of
> Balliol College. I caught cold and was required to stay in
> bed for about three days. After two days I had got tired of
> reading detective stories and science fiction, and I began
> thinking about the structure of proteins.

Pauling started his new onslaught on the puzzler with the
assumption that all the amino acids in the alpha keratin should be
in a structurally similar position with respect to the polypeptide
chain. While still in bed, he asked his wife, Ava Helen, to bring him
a pencil, a ruler, and a piece of paper. Keeping each peptide group in
the plane of the paper, using heavier and lighter lines to indicate the
three-dimensional relationships, and rotating around the two single bonds to the carbon atoms (with the angle of rotation being the
same from one peptide group to the next), Pauling created a *helix*,
a spiral-staircase-like structure, in which the polypeptide backbone
formed the core of the helix, and the amino acids projected outward
(figure 12). To stabilize the construction, Pauling formed hydrogen
bonds between one turn of the helix and the next turn, parallel to the
helix axis. (Figure 12; a hydrogen bond is a chemical bond in which
a hydrogen atom of one molecule is attracted to an atom of another
molecule.) He actually found two structures that could work, one of
which he called the alpha-helix, and the other the gamma-helix. That
Pauling was able to find solutions to the problem with such primitive tools attests to how crucial his previous discovery of the planarity of the peptide group had been. (Figure 11 represents his attempt
to reconstruct the original piece of paper from 1948.) Without it,
the number of possible conformations would have been much larger.

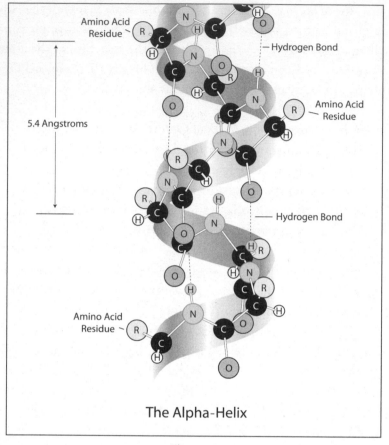

The Alpha-Helix

Figure 12

Excited, Pauling asked his wife to bring him a slide rule (long obsolete, this was the most commonly used calculation tool at the time), so that he could calculate the repeat distance along the fiber axis. He discovered that the structure of the alpha-helix was repeating after 18 amino acids in five turns. That is, the alpha-helix had 3.6 amino acids per turn. Alas, to his disappointment, the calculated distance between turns was 5.4 angstroms, and not the 5.1 angstroms hinted at by the X-ray diffraction patterns. The gamma-helix had a hole down its center that was too small to be occupied by other molecules, so Pauling concentrated his attention on the alpha-helix. Feeling fairly confident in the correctness of his solution, Pauling tried very hard to find some way to adjust either the bond lengths or the bond angles so as to decrease the calculated distance from 5.4 to 5.1 angstroms, but he failed to do so. Consequently, even though he was extremely pleased with his alpha-helix, he decided to refrain from publishing the model until he could understand better the reason for the discrepancy in the spacing.

About six weeks later, Pauling visited the Cavendish Laboratory at Cambridge, and what he saw there impressed him deeply. "They have about five times as great an outfit as ours," he wrote to his assistant at Caltech, "with facilities for taking nearly 30 X-ray pictures at the same time." Concerned that there was something still wrong with his model, and at the same time anxious that the Cavendish group might beat him to analyzing it, Pauling remained silent about the alpha-helix. Even during a discussion with the famous chemist Max Perutz, in which the latter showed him exciting new results on the structure of the hemoglobin crystal, Pauling decided to keep his ideas to himself.

The problem, however, continued to haunt him. Upon his return to Pasadena, Pauling immediately asked a visiting professor of physics, Herman Branson, to inspect his calculations carefully. Pauling was particularly interested to know whether Branson could find a third helical structure that would satisfy the restrictions of a planar peptide bond and maximum hydrogen bonding for stability. Branson and one of Pauling's research assistants, Sidney Weinbaum, went over Pauling's computations with a fine-tooth comb for about a year and concluded that there were truly only two structures—the alpha-helix and the gamma-helix—that satisfied all the constraints. Branson and Weinbaum also confirmed that the alpha-helix, which was the tighter of the two helices, was characterized by a distance of 5.4 angstroms between turns.

Pauling was now presented with the choice of whether simply to ignore the incongruity with the X-ray data and to publish his model or to delay publication until that conundrum was resolved fully. A paper submitted for publication to the *Proceedings of the Royal Society of London* on March 31, 1950, helped him decide.

I Wish I Had Made You Angry Earlier

The paper, entitled "Polypeptide Chain Configurations in Crystalline Proteins," was written by an illustrious trio: Lawrence Bragg, Nobel Prize for Physics laureate in 1915, and two molecular biologists who eventually shared the 1962 Nobel Prize for Chemistry—John Kendrew and Max Perutz—all from the Cavendish Laboratory at Cambridge. At the time, this famous lab was the world's leading center for X-ray crystallography. This method for analyzing crystals was largely Bragg's baby; he and his father, Sir William Henry Bragg, together worked out the mathematics underlying the physical phenomenon and developed the experimental technique.

The idea behind X-ray crystallography was genius in its simplicity. Physicists had known from the beginning of the nineteenth century that if they shined visible light onto a finely spaced grating, the light that passed through formed a diffraction pattern of bright and dark spots on a screen on the other side. The bright spots marked the locations where light waves from different slits in the grating combined to enhance each other, while the dark spots formed where the different waves underwent destructive interference (as when a crest from one wave was superimposed onto a trough from another). Physicists also knew, however, that for this diffraction pattern to form, the spacings between the different slits needed to be of the same order as the wavelength of the light (the distance between two successive crests in the wave). While it was relatively easy to manufacture such fine gratings for visible light, it was impossible to produce them for X-rays, since a typical wavelength for X-rays is a few thousand times shorter than wavelengths in the visible part of the spectrum. The first person to realize that natural, periodic crystals could serve as gratings for X-ray diffraction experiments was Max von Laue. The German physicist recognized that the inter-atomic distances in crystals were precisely of the order of the presumed wavelengths of X-rays. Following in Laue's footsteps, Lawrence Bragg formulated the mathematical law describing the diffraction of X-rays on a crystalline structure. Amazingly, he obtained this important result during his first year as a research student at Cambridge. The father and son team then went on to build the X-ray spectrometer that allowed them to analyze the structure of many crystals. Lawrence Bragg remains, by the way, the youngest person to be awarded a Nobel Prize. (He won it at age twenty-five!)

Given this formidable legacy, we can imagine that when Pauling saw the title of the paper by Bragg, Kendrew, and Perutz, his heart missed a beat. The first two paragraphs of the paper indeed gave the impression that Bragg's team may have beaten him to the punch: "Proteins are built of long chains of amino-acid residues . . . In this paper an attempt is made to glean as much information as possible about the nature of the chain from x-ray studies of crystalline proteins, and to survey the possible types of chain which are consistent with such evidence as is available." Pauling read quickly all thirty-seven pages and was relieved to discover that while the Cavendish researchers described some twenty structures, the alpha-helix wasn't one of them. Moreover, they concluded that none of the examined structures was acceptable as a model for alpha keratin. Pauling agreed happily with this conclusion, especially since he thought that Bragg's team did not apply the most important constraint to its configurations but did impose a handicap that he regarded as totally unnecessary. On one hand, none of Bragg's models assumed the planarity of the peptide group, of whose correctness Pauling was absolutely convinced. On the other, the Cavendish team appeared to be hung up on the notion that in every full turn of its helical structures, there had to be an *integer number* of amino acids. Pauling's alpha-helix broke with tradition and had about 3.6 amino acids per turn, and he saw nothing wrong with that. Coming from an X-ray crystallography background, Bragg also adhered religiously to the apparent 5.1 angstrom distance between turns suggested by Astbury's data. Perutz later described that to start the team off, Bragg hammered nails representing amino acid residues into a broomstick in a helical pattern with an axial distance between successive turns of 5.1 centimeters.

Pauling was always extremely competitive in nature. Even though he was pleased to see that the Cambridge team had missed a few key points, the appearance of Bragg's paper prompted him into action, for fear he might be scooped. In October 1950 he and Corey sent a short note describing the alpha-helix and the gamma-helix to the *Journal of the American Chemical Society.* Around the same time, some encouraging results were coming from another British research group at Courtaulds Research Laboratories. There, Clement Bamford, Arthur Elliott, and their collaborators succeeded in producing fibers of synthetic polypeptides. To Pauling's delight, X-ray diffraction photographs of those fibers showed clearly that the repeat distance along the axis was 5.4 angstroms—consistent with Pauling's findings—rather than 5.1 angstroms. This raised the suspicion that the latter feature in the X-ray photographs of hair could simply be an artifact produced by overlapping reflections rather than a major clue to the structure. Increasingly convinced of the truth of this interpretation, Pauling submitted a paper by himself, Corey, and Branson that contained a detailed explanation of the alpha- and gamma-helices. It was only fitting that this important paper was submitted precisely on the day of Pauling's fiftieth birthday, February 28, 1951.

There is, incidentally, an interesting anecdote concerning the use of the term "helix," which I heard from chemist Jack Dunitz, who at the time was a postdoctoral fellow with Pauling. Dunitz recalled that in 1950 Pauling kept using the term "spiral" to describe the structure of alpha keratin. Even in Pauling and Corey's short communication in the *Journal of the American Chemical Society*, they wrote exclusively about spirals. One day, said Dunitz, he remarked to Pauling that he thought that the word "spiral" referred only to the two-dimensional, planar shape, while the three-dimensional one had to be called a "helix." Pauling responded that a spiral could be either two-dimensional or three-dimensional, but added that on second thought, he liked the word "helix" better. When the extensive manuscript by Pauling, Corey, and Branson was submitted, it avoided the word "spiral" altogether. Its title read: "The Structure of Proteins: Two Hydrogen-Bonded Helical Configurations of the Polypeptide Chain." Pauling was by then so confident in his model that he and

Corey followed the alpha-helix paper with a barrage of papers on the folding of polypeptide chains.

That spring in England, Max Perutz went one Saturday morning to the library, and there, in the latest issue of the *Proceedings of the National Academy of Sciences,* he found the series of papers by Pauling. Some thirty-six years later, he described what he had experienced that morning (in a somewhat technical language, but the emotions were crystal clear):

> I was thunderstruck by Pauling and Corey's paper. In contrast to Kendrew's and my helices, theirs was free of strain; all the amide groups were planar and every carboxyl group formed a perfect hydrogen bond with an amino group four residues further along the chain. The structure looked dead right. How could I have missed it? Why had I not kept the amide groups planar? Why had I stuck blindly to Astbury's 5.1 angstrom repeat? On the other hand, how could Pauling and Corey's helix be right, however nice it looked, if it had the wrong repeat? My mind was in a turmoil. I cycled home to lunch and ate it oblivious of my children's chatter and unresponsive to my wife's inquiries as to what the matter was with me today.

Thinking a bit more about Pauling's model, Perutz realized that the alpha-helix resembled a helical staircase, in which the amino acid residues (marked by "R" in figure 12) were forming the "steps." The height of each step was about 1.5 angstroms. Bragg's X-ray diffraction theory therefore predicted the existence of never-before-reported X-ray reflection signatures, separated by 1.5 angstroms, from planes perpendicular to the fiber axis. None of the models of Bragg's group would have produced such a mark, while this would have been a distinct "fingerprint" of Pauling's alpha-helix.

Just as he was about to conclude that the lack of such reflections in Astbury's data was sufficient to refute Pauling's model, Perutz suddenly recalled that Astbury's particular experimental setup—

with the fibers oriented such that their long axes were perpendicular to the beam of X-rays—would not have really allowed for the detection of the 1.5 angstrom signature. Rather, calculations predicted that the optimal conditions to observe the reflection would have required inclining the fibers at an angle of about 31 degrees.

Perutz felt absolutely compelled to make the crucial test right away. He cycled back to the lab, grabbed a horsehair he had in a drawer, inserted it into the apparatus at the angle he calculated to be favorable for detecting the reflection, put a film around it (as opposed to Astbury's flat-plate camera, which was too narrow and could have missed reflections deflected at large angles), and fired the X-ray beam. The few hours that passed before he could develop the film were sheer agony, but finally Perutz had the answer. The strong reflection predicted by the alpha-helix at a spacing of 1.5 angstroms stuck out unambiguously!

Perutz showed the X-ray photograph to Bragg first thing on Monday morning. Bragg wondered what it was that suddenly gave Perutz the idea to conduct this crucial test. Perutz replied that he was madly furious with himself for not having thought of the alpha-helix. Bragg retorted with what has by now become an immortal phrase: "I wish I had made you angry earlier!"

Life's Blueprint

Not everything that Pauling wrote in that famous series of papers from 1951 was correct. A careful scrutiny of his entire oeuvre for that year reveals several weaknesses. In particular, the gamma-helix eventually had to be abandoned. These minor shortcomings, however, don't take away anything from Pauling's groundbreaking achievement: the alpha-helix and its prominent role in the structure of proteins. Pauling's contributions to our understanding of the nature of life were substantial. He was one of the first scientists to see that in spite of its inherent complexity, biology is, at its core, molecular science augmented by the theory of evolution. Already back in 1948, he wrote perceptively: "To understand all these great biological phenomena we need to understand atoms, and the molecules that they form by bonding together; and we must not be satisfied with an understanding of simple molecules . . . We must also learn about the structure of the giant molecules in living organisms."

Pauling's influence on the general theory and methodology of molecular biology was equally impressive. First, in his seminal 1939 book *The Nature of the Chemical Bond and the Structure of Molecules and Crystals: An Introduction to Modern Structural Chemistry*, he remarked prophetically on the importance of the hydrogen bond for biomolecules: "I believe that as the methods of structural chemistry are further applied to physiological problems it will be found that the significance of the hydrogen bond for physiology is greater than that of any other single structural feature." Indeed, the structure of many organic molecules, ranging from proteins to nucleic acids, confirmed this prediction fully.

Second, Pauling pioneered model building and turned it into a predictive art form based on strict rules of structural chemistry. Even the space-filling colored models developed at Caltech became a hot item in the arena of macromolecular research. These models, produced for labs by the Caltech workshop, fetched as much as $1,220 in 1956 for a set that contained about six hundred atom models.

Pauling's practice of using the X-ray diffraction patterns not as the starting point but as the ultimate arbiter among sophisticated, educated guesses also proved to be enormously effective—Watson and Crick were about to apply the same approach to the structure of DNA.

There was another remarkable observation concerning genetics that Pauling made in a lecture in 1948, but apparently even he did not realize at the time its full implications. In the first part of that lecture, Pauling reminded his audience:

> The Gregorian monk Mendel noted that the inheritance of characters by pea plants, such as the character of tallness or of dwarfness, or the character of having purple flowers or white flowers, could be understood on the basis of

hereditary units transmitted from the parent to the off-spring. Thomas Hunt Morgan and his collaborators identified these units with genes arranged in a linear array in the chromosomes.

Then, toward the end of the lecture, he added the following comment:

> The detailed mechanism by means of which a gene or a virus molecule produces replicas of itself is not yet known. In general the use of a gene or virus as a template would lead to the formation of a molecule not with identical structure but with complementary structure. It might happen, of course, that a molecule could be at the same time identical with and complementary to the template on which it is moulded. However, this case seems to me to be too unlikely to be valid in general, except in the following way. *If the structure that serves as a template (the gene or virus molecule) consists of, say, two parts, which are themselves complementary in structure, then each of these parts can serve as the mould for the production of a replica of the other part, and the complex of two complementary parts thus can serve as the mould for the production of duplicates of itself* [emphasis added].

As we shall soon see, had Pauling remembered his own pronouncement four years later when he was trying to determine the structure of DNA, he might have avoided making a terrible blunder.

Pauling started to turn his attention to DNA only in the summer of 1951. Until the early 1950s, most life scientists subscribed to the protein paradigm: the view that proteins, rather than nucleic acids, formed the foundation for life and were the crucial players in reproduction, growth, and regulation. The roots of this view could be traced to biologist Thomas H. Huxley ("Darwin's Bulldog"), who believed that the protoplasm—the living part of the cell—was the source of all of life's attributes. Proteins, which are built up of amino acids in a long chain, make up a large fraction of all living cells, while nucleic acids, as their name implies, were found first in the nuclei of cells.

The early work on the structure and constitution of the nucleic acids by biochemist Phoebus Levene did not help to spark interest in these molecules. If anything, his studies achieved precisely the opposite effect. Levene managed to distinguish the deoxyribonucleic acid (DNA) from ribonucleic acid (RNA), and to find some of their properties. But his results generated the impression that these were rather simple and dull substances unsuited for the complex tasks of governing growth and replication. In the words of cytologist Edmund Beecher Wilson (in 1925): "The nucleic acids of the nucleus are on the whole remarkably uniform . . . In this respect they show a remarkable contrast to the proteins, which, whether simple or compound, seem to be of inexhaustible variety." This impression persisted throughout the 1940s. By then, DNA was known to be composed of unbranched chains of units called *nucleotides.* The nucleotides themselves also appeared to be fairly uncomplicated, with each one containing three subunits: a *phosphate* group (a phosphorus atom bonded to four oxygen atoms), a five-carbon *sugar,* and one of four nitrogen-containing *bases.* The four bases were: *cytosine* and *thymine,* which were single ringed; and *adenine* and *guanine,* which were both double ringed (see figure 13). What was still not known, even in 1951, was the actual structure: how exactly the subunits connected to each other to form nucleotides, and the nature of the links between the nucleotides themselves. However, while all of this seemed to be fairly interesting from a chemical perspective, at the end of 1951 most geneticists still believed that DNA's only role was structural, acting perhaps as a scaffold for the more sophisticated proteins rather than being directly related to heredity.

This fact in itself was somewhat surprising, given that in a paper published back in 1944, biologists Oswald Avery, Colin MacLeod, and Maclyn McCarty provided strong experimental evidence that the genetic material of living cells was composed of DNA. Avery

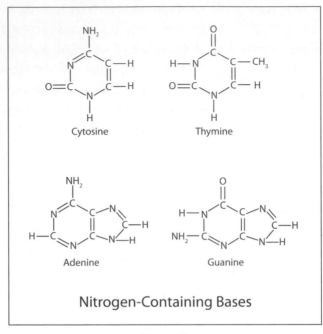

Nitrogen-Containing Bases

Figure 13

and his colleagues grew large quantities of virulent bacteria, and after managing to separate them into their biochemical constituents, they concluded that DNA molecules—and not proteins or fats—were the components responsible for converting nonvirulent bacteria into virulent ones. In a May 1943 letter describing the results to his bacteriologist brother, Roy, Avery concluded, "So there's the story, Roy—right or wrong it's been good fun and lots of work." The reason that Avery's findings did not get the attention they deserved may have had to do with the fact that since none of the three scientists was a geneticist, their conclusions were formulated with such caution that many of the life scientists failed to appreciate their full import. The statement in the paper read: "If it is ultimately proved beyond reasonable doubt that the transforming activity of the material described is actually an inherent property of the nucleic acid, one must still account on a chemical basis for the biological specificity of its action." Still, careful readers should have taken notice of the paper's summary: "The data obtained . . . indicate that, within the limits of the methods, the active fraction contains no demonstrable protein . . . and consists principally, if not solely, of a highly polymerized, viscous form of desoxyribonucleic acid [DNA]."

Pauling was familiar with Avery's work, but even he admitted in a later interview that at the time he did not believe that DNA had much to do with heredity: "I knew the contention that DNA was the hereditary material. But I didn't accept it; I was so pleased with proteins, you know, that I thought that proteins probably are the hereditary material, rather than nucleic acid." Chemist Peter Pauling, Linus's son, also affirmed that this had indeed been his father's attitude. In a short article written in 1973, Peter reported, "To my father, nucleic acids were interesting chemicals, just as sodium chloride [ordinary table salt] is an interesting chemical, and both presented interesting structural problems."

Nevertheless, toward the end of 1951, an unusual paper by biochemist Edward Ronwin, then at the University of California at Berkeley, intrigued Pauling sufficiently to prod him into action. The paper, entitled "A Phospho-tri-anhydride Formula for the Nucleic Acids," appeared in November 1951. In it, Ronwin proposed a new "design" for DNA, in which each phosphorus atom connected to five oxygen atoms, while Pauling—the consummate structural chemist—was absolutely convinced that it had to link only to four. Annoyed, Pauling fired a quick communication to the editor of the *Journal of the American Chemical Society* (together with chemist Verner Schomaker) in which they first noted that "in formulating a hypothetical structure for a substance, one must take care that the structural elements of which use is made are reasonable ones." Their conclusion was even more dismissive: "The ligation of five oxygen atoms about each phosphorus atom is such an unlikely structural feature," they said, that the proposed formula for DNA "deserves no serious consideration." Ronwin retorted by pointing out that other substances in which phosphorus was bonded to five oxygen atoms did exist. Pauling and Schomaker had to withdraw their disparaging statement, but they still insisted correctly on the fact that structures of this type were extremely sensitive to moisture, which made them unlikely candidates for DNA. This exchange would have been insignificant except that it did get Pauling thinking about how DNA might be constructed. To make progress, however, he needed high-quality X-ray diffraction photographs of DNA, since the ones available in print were old photos taken by William Astbury and Florence Bell in 1938 and 1939. Unfortunately, good X-ray photos were not easy to come by. Caltech did produce new photographs in the early 1950s, but, surprisingly, those turned out to be of inferior quality to those of Astbury and Bell. While weighing his options, Pauling heard that Maurice Wilkins of King's College, London, had generated what were described as "good fibre pictures of nucleic acid." Deciding that he had nothing to lose, Pauling wrote to Wilkins to inquire whether the latter was prepared to share those photos. Unbeknown to Pauling, however, the activity around DNA in England was rapidly approaching frenzy.

Meanwhile, in England

Three separate events, all happening in 1951, proved to be fateful for the "race" to uncover the structure of DNA. In that year, Francis Crick, at age thirty-five, was working at Cambridge toward a PhD degree in biology after having been bored with physics. (He later described his work on the viscosity of water as "the dullest problem imaginable.") His mathematical background would be crucial for the discoveries to come. In the same year, James Watson, then twenty-three, arrived at Cambridge to learn about X-ray diffraction from Max Perutz. Watson had completed his PhD at the University of Indiana on the effects of X-rays on viruses and later had some training in nucleic acid chemistry at the University of Copenhagen. Also in 1951, Rosalind Franklin, then thirty-one, came to King's College, after having completed three years of research in Paris, where she became proficient in X-ray diffraction techniques.

Franklin, who came from an erudite banking family, had earned her PhD from Cambridge in 1945. When she arrived at King's College, physicist Maurice Wilkins was hoping that by virtue of her being an accomplished crystallographer, she would help him in his studies of molecular structure. That Wilkins would expect that of Franklin was not at all surprising, since at the time, according to Watson's account, "molecular work on DNA in England was, for all practical purposes, the personal property of Maurice Wilkins." This was not at all, however, what Franklin had in mind when she signed up to come to King's, and she had good reasons for her different presumption. Sir John Randall, director of the college's Biophysics Research Unit, had written her a letter in which he described her job as follows: "This means that as far as the experimental X-ray effort is concerned there will be at the moment only yourself and Gosling [Raymond Gosling, who was a graduate student at the time], together with the temporary assistance of a graduate from Syracuse, Mrs. Heller." Franklin was therefore under the logical impression that she was going to be her own boss as far as the DNA work was concerned—an attitude that clearly conflicted with Wilkins's assumptions. Consequently, Franklin and Wilkins were bound to clash, and they did. Later, they ended up working separately, even though they shared the same laboratory quarters.

By contrast, Watson and Crick, who were sharing an office at Cambridge, hit it off right away. Watson described Crick as "no doubt the brightest person I have ever worked with and the nearest approach to Pauling I have ever seen." The two men brought together rather different but complementary expertise, traits, and temperaments. As Crick noted in an interview, "The interest was that his [Watson's] background was in phage work which I had only read about and did not know first hand and my background was in crystallography which he had only read about and did not know first hand." It is amusing to read how they described each other's personality. Referring to Crick's assuredness, puckish wit, and habit to speak his mind, Watson wrote about him, "I have never seen Francis Crick in a modest mood." He also added that Crick "talked louder and faster than anyone else." On the other hand, Crick wrote about

Watson, "Jim was distinctly more outspoken than I was." Despite the different backgrounds, something clicked immediately between the two. Crick suspected that this was "because a certain youthful arrogance, a ruthlessness, and an impatience with sloppy thinking" came naturally to both of them. Their thought processes were also fairly similar. In Crick's words, "He was the first person I had met who thought the same way about biology as I did . . . I decided that genetics was the really essential part, what the genes were and what they did."

There was something else that made the Watson-Crick collaboration truly powerful. Because neither of them was professionally senior to the other, they could afford to be brutally honest in criticizing each other's ideas. This type of intellectual honesty is sometimes missing in relationships burdened by formal politeness, bowing to one's superiority, or by one or the other pulling rank. This is how Crick himself described his interaction with Watson: "If either of us suggested a new idea, the other, while taking it seriously, would attempt to demolish it in a candid but nonhostile manner." According to Crick, Watson "was determined to discover what genes were and hoped that solving the structure of DNA might help." This turned out to be absolutely true.

One may still wonder what it was that convinced Watson and Crick that the DNA structure was at all solvable rather than being an irregular mess. Conceivably, it was a talk Maurice Wilkins had given at a meeting in Naples, Italy, in the spring of 1951—a meeting that Watson attended. Wilkins succeeded in pulling extremely thin fibers of the sodium salt of DNA, and in producing X-ray photographs that were significantly superior to the ones by Astbury and Bell. The pictures showed a crystalline form of DNA, which indicated to Watson that the structure was regular. These were the same pictures that Pauling had requested from Wilkins.

Upon receiving Pauling's letter, Wilkins, who was fully aware of Pauling's abilities when it came to molecular structure, did not quite know what to do about the request. Eventually, he replied politely that his pictures were not ready to share until he had the opportunity to carry out some additional investigations. Pauling did not give up, and he decided to try his luck with Randall, only to be refused again on the grounds that "it would not be fair to them [Wilkins and his collaborators], or to the efforts of our laboratory as a whole, to hand these over to you." So by the end of 1951, Pauling was still unable to see any X-ray diffraction images of reasonable quality.

Meanwhile, Watson and Crick were becoming increasingly obsessed with the desire to beat Pauling at deciphering the structure of DNA. The Austrian-American biochemist Erwin Chargaff, who met Watson and Crick in May 1952, gave a humorous description of the dynamic duo: "One, thirty-five years old; the looks of a fading racing tout, something out of Hogarth . . . The other, quite undeveloped at twenty-three, a grin, more sly than sheepish; saying little, nothing of consequence." Even funnier was Chargaff's depiction of the burning ambition of the two scientists: "So far as I could make out, they wanted, unencumbered by any knowledge of the chemistry involved, to fit DNA into a helix. The main reason seemed to be Pauling's alpha-helix model of a protein." Indeed, even though Pauling was unaware of it, Watson (in particular) and Crick (to some extent) saw themselves as participating in a race against him.

One should not get the impression that Pauling was the first to introduce helical models, but he certainly had a major role in making such models the choice for molecules of biological significance. By introducing a non-integer number of amino acids per turn in his alpha-helix model, Pauling expanded further the horizons of traditional structural crystallographers. Consequently, research into the interpretation of the X-ray diffraction patterns from helical structures received a huge boost, establishing the tools for the eventual deciphering of DNA. As Crick described the general thinking around that time: "You would be eccentric, looking back, if you didn't think DNA was helical."

Toward the end of 1951, events started to progress rapidly. On November 21, 1951, Watson made a trip to London to hear a colloquium by Rosalind Franklin. Even though he did not learn much new from that lecture, barely a week passed before he and Crick produced their first model for the structure of DNA. The model consisted of three helical strands and had a sugar-phosphate backbone on the inside, with the bases pointing outward. The main motivation for this particular design was simple: Since the bases were of different sizes and shapes (two were single ringed and two were double ringed; see figure 13), Watson and Crick didn't see how the crystalline DNA could produce a highly regular pattern unless the bases were relatively uninvolved in the central architecture.

On John Kendrew's advice, the energized pair invited the King's College team to see their model, even though Crick later admitted that he had felt somewhat uncomfortable to issue the invitation so soon. The overture was accepted immediately: The group of Maurice Wilkins, Rosalind Franklin, Raymond Gosling, and William Seeds (another member of the Biophysics Research Unit) showed up in Cambridge the very next day.

The presentation of the first Watson and Crick model proved to be a total disaster. Not only did Franklin question all of the basic assumptions, from the helical structure to the forces that were supposed to hold together the core, but she also pointed out that the reported water content was completely wrong—DNA was a rather "thirsty" molecule—invalidating all of Watson's density calculations. Apparently, part of the mistake was due to Watson's misunderstanding of a crystallographic term that Franklin had used in her seminar a week earlier. This unfortunate confusion led Crick to believe that the number of possible configurations was rather limited.

The fiasco had meaningful consequences: Watson and Crick were essentially banned from continuing their DNA work, and all the DNA research was supposed to be confined exclusively to King's College in London. It has usually been assumed that the directors of the two labs, Randall and Bragg, called for the moratorium on additional DNA work by Watson and Crick. However, in 2010 Alexander Gann and Jan Witkowski of the Cold Spring Harbor Laboratory in New York discovered some long-lost correspondence of Francis Crick's. As it turned out, the missing letters had become mixed with papers of biologist Sydney Brenner, with whom Crick shared an office between 1956 and 1977. The recovered correspondence provides a new perspective on the circumstances of the suspension of the DNA research. In a formal letter dated December 11, 1951, Maurice Wilkins wrote to Crick:

> I am afraid the average vote of opinion here [at King's College], most reluctantly and with many regrets, is against your proposal to continue to work on n. a. [nucleic acids] in Cambridge. An argument here is put forward to show that your ideas are derived directly from statements made in a colloquium and this seems to me as convincing as your own argument that your approach is quite out of the blue.

Wilkins, continuing to assume the role of the mediator between King's and the Cavendish, then added, "I think it most important that an understanding be reached such that all members of our laboratory can feel in future, as in the past, free to discuss their work and interchange ideas with you and your laboratory. We are two M. R. C. [Medical Research Council] Units and two Physics Departments with many connections." Wilkins suggested further that Crick should show the letter to Max Perutz, and he informed Crick that he was giving a copy to Randall. On the same day, Wilkins also sent Crick a more personal, handwritten letter in which he confessed that he "had to restrain Randall from writing to Bragg complaining about your behavior." A draft reply written by Watson and Crick two days later indicates that "we've all agreed that we must come to an amicable arrangement." Watson however, was not going to be deterred from at least cogitating over DNA by an administrative decision.

Meanwhile, Franklin, on her part, was making significant progress. First, she discovered that DNA occurred in two somewhat different configurations. One form, which she labeled "A," was crystalline. The other, the "B" form, was more extended and contained more water. One of the consequences of the existence of these two conformations was that the X-ray diffraction photos of DNA samples looked confused unless produced from one pure form. Franklin spent the first five months of 1952 generating pure samples of both the A and B forms, managing to pull out single fibers of each form, and designing and reconfiguring her X-ray camera to take high-resolution pictures. As we shall see shortly, one of the photographs she had produced of the "wetter" B form, which was tagged photograph 51 (see figure 14), was about to become key for understanding the structure of DNA. Unfortunately, because Franklin decided to use a particular method of analysis, she and Gosling concentrated first on the more detailed X-ray pictures of the A form, neglecting the simpler but truly revealing X-ray pattern in photograph 51 for almost nine months!

In all of her research endeavors, Franklin exhibited a striking difference between her way of thinking and Pauling's. Franklin abhorred "educated guesses" and heuristic methods. Rather, she insisted on relying on the X-ray data to lead her to the right answer. For instance, although she did not object in principle to helical structures, she absolutely refused to *assume* their existence as a work-

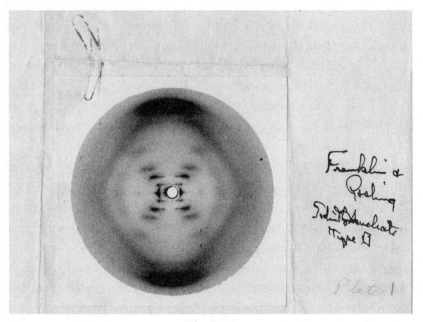

Figure 14

ing hypothesis. In contrast, Watson and Crick emulated Pauling's approach and methods to the fullest, and they were not going to be bogged down in formal methodology. In Crick's words, "He [Watson] just wanted the answer, and whether he got it by sound methods or flashy ones did not bother him one bit. All he wanted was to get it as quickly as possible."

Surprisingly, neither Watson and Crick nor Pauling knew at the time that already in 1951, Elwyn Beighton in Astbury's lab at the University of Leeds had produced excellent X-ray photographs of the B form by stretching the DNA fibers and wetting them. However, since Astbury and Beighton apparently thought that this represented a mixture rather than a pure configuration (because the X-ray pattern was simpler than in the Astbury-Bell pictures), they did not advertise the existence of these photos at all. Unfortunately for Astbury and Beighton, neither of them was familiar with how a helix would appear in X-ray photographs. Just like that, the Leeds lab missed an opportunity to play a significant role in the DNA story.

Back in the United States, Pauling was trying to work his magic again with DNA, in an attempt to replicate his feat with proteins. The available X-ray photos showed a strong reflection at about 3.4 angstroms, but nothing much else. As a starting point, Pauling reexamined Ronwin's paper. Even though he was convinced that Ronwin's proposed structure for the DNA, with the phosphorus atom connected to five oxygen atoms, was completely wrong, something in Ronwin's suggestion attracted his attention. Ronwin had the four bases on the outside of the structure and the phosphates down the middle. This seemed to make sense to Pauling, precisely for the same reason that Watson and Crick placed the bases on the outside in their first attempt. (Pauling was unaware of that totally off-target model.) Following this line of thought, Pauling embarked again on what has become known as his "stochastic method." The idea was to use chemical principles to pare down the list of possible structures to the most plausible ones and then to construct three-dimensional models of those in order to eliminate configurations that were packed either too tightly or too loosely. He could then check the emerging "best-bet" arrangement against the experimental X-ray diffraction pattern.

Having had great success with this method on previous occasions, Pauling thought that he knew exactly which steps to follow. First, he had little doubt that the molecule was helical, and the Astbury-Bell photographs seemed to be generally consistent with this assumption. Second, two of the bases were double ringed and two were single ringed. The different constructions and dimensions made it difficult, at least at first glance, to have the core of a helix—which appeared to be regular—be composed of the bases. The next step was to figure out how many strands the helix should have. Pauling decided to attack this problem by calculating the density of the structure. However, before he was even able to start, an unexpected distraction stopped him in his tracks.

Life Under McCarthyism

In the Cold War atmosphere that followed World War II, and in particular after the passage of the Internal Security Act of 1950, the US State Department's Passport Division was given almost unlimited authority to deny passports to anyone it deemed to be too "leftist." Pauling applied to renew his passport in January 1952, as he was preparing to attend a Royal Society meeting in London the following May. Pauling and Corey had both been invited to present their work on proteins and the alpha-helix at that conference, and Pauling was also planning to take advantage of this trip to Europe to visit a few universities in Spain and France. Then on February 14, 1952, Ruth B. Shipley, head of the Passport Division, sent Pauling a letter that could hardly be considered a Valentine's Day card. She informed him that his passport could not be issued, since the department was of the opinion that his travel "would not be in the best interest of the United States."

In the then-prevailing mood, given Pauling's many pacifist speeches, his activism against nuclear weapons, and his declaration that "the world now stands at a branch in the road, leading to a glorious future for all humanity or to the complete destruction of civilization," it was perhaps not entirely shocking that Shipley would surmise that "there is good reason to believe that Dr. Pauling is a Communist."

At first, Pauling regarded the refusal merely as an annoying inconvenience, and he was convinced that the problem would be resolved easily. To speed things up, he immediately sent a letter to President Harry Truman, to which he attached a copy of his 1948 Presidential Medal for Merit, signed by Truman. Pauling wrote in frustration, "I am confident that no harm whatever would be done to the Nation by my proposed travel." The president's secretary replied politely that the Passport Division had been asked to reevaluate its judgment. Still, the decision was not reversed. In April, with an increasing sense of urgency, Pauling took a series of actions: First, he asked for the assistance of a lawyer. Second, he supplied the Passport Office with loyalty oaths and affidavits declaring he was not a Communist. Finally, he arranged to meet in person with Ruth Shipley. All of this was to no avail. The conclusive denial of the appeal was announced on April 28, and the next day, Pauling notified the organizers of the Royal Society meeting of his inability to attend.

Predictably, Pauling's passport trials and tribulations infuriated scientists worldwide. Sir Robert Robinson, the Nobel laureate chemist from England, wrote a letter to the London *Times* expressing his "consternation." Leading American and British scientists, including physicists Enrico Fermi and Edward Teller, biologist Harold Urey, and crystallographer John Bernal, wrote letters in protest, and French biochemists elected Pauling to be the honorary president of an International Biochemical Congress scheduled to take place in Paris in July.

The international pressure eventually had an effect. When Pauling reapplied for a passport in June, the State Department overturned Shipley's denial, and Pauling was allowed, on July 14 (Bastille Day), to travel to France and England.

In addition to its political significance, the entire passport debacle had some scientific consequences as well. Corey, who did attend the Royal Society meeting, used the opportunity to visit Franklin's laboratory. There he was shown the superb X-ray photos she had obtained. However, he apparently did not grasp immediately the full implications of the photographs, since he did not communicate anything of significance to Pauling. Volumes of speculation have been written about what might have happened had Pauling himself been allowed to travel to see those photographs. These speculations are, in fact, quite irrelevant. Pauling *had* every opportunity to visit the King's College team just ten weeks later, during the month he spent in England in the summer of 1952, and he chose not to do so. The reason was simple: Pauling was still focused on convincing everybody about the correctness of his alpha-helix model for proteins; DNA was not the main topic on his mind. As it later transpired, Franklin's photos—in particular the soon-to-become-famous 51—contained the clear hallmarks of a double-stranded helix.

There was yet another important piece of information concerning DNA that Pauling had been made aware of but either had forgotten or at least had not internalized. This evidence was related to the bases in the nucleotides. The following anecdote demonstrates how emotional responses may interfere even with processes that are supposed to be governed by pure scientific reasoning.

The day after Christmas 1947, Pauling and his family had been on their way to Europe for Pauling's six-month visit to Oxford. They traveled on board the famous *Queen Mary*. Coincidentally, Erwin Chargaff, who had been interested in nucleic acids since the war years, happened to be on board the same ship, and Pauling soon ran into him. Unfortunately, Chargaff was, in the words of biologist Alex Rich, a "very intense individual." This did not suit Pauling, who was generally easygoing and, in this particular instance, was looking forward to a relaxing vacation. Consequently, Pauling not only paid little attention to Chargaff's animated description of his research results but also later seemed to have ignored Chargaff's important paper on nucleic acids. In that paper, published in 1950, Chargaff discovered a remarkable relation between the amounts of the bases in DNA. He showed that whatever the number of adenine molecules (usually abbreviated "A") in a certain section of DNA, the number of thymine (abbreviated "T") molecules was equal.

Similarly, the number of guanine units ("G") was equal to the number of cytosine ("C") units. This meaningful clue to the structure of DNA—that the amount of A is equal to the amount of T, and the amount of G equals the amount of C—completely escaped Pauling's attention. If it hadn't, perhaps the discovery of DNA's structure would have played out differently.

Following his trip to England and France in the summer of 1952, Pauling returned to Caltech in September. However, even then he was not ready to plunge back fully into the DNA problem. A conversation he had with Crick in England that summer gave him an idea of how he could finally resolve the puzzle of the protein reflection at 5.1 angstroms. As often happens in science, Pauling and Crick solved that problem independently, each showing that the alpha-helices could form coiled, ropelike structures around one another, those giving rise to the enigmatic signature. This had a nice ring of finality to it, but even though Pauling didn't know it at the time, the "race" to solving DNA was entering the homestretch.

The Triple Helix

Pauling's visit to France provided him with an additional clue to the fact that it was probably DNA, after all, that was the primary genetic material. The American microbiologist Alfred Hershey presented the evidence in a talk at an international meeting on viruses in Royaumont, near Paris. Hershey and his collaborator, Martha Chase, labeled the DNA and protein of the T2 phage (a virus) with radioactive phosphorus and sulphur, respectively. They then allowed the phages to infect bacteria, and were able to demonstrate that the genetic material that infected the bacteria was most probably DNA and not protein. The viral protein coat remained outside the bacterial cell and played no role in the infection. But not everyone was convinced. Indeed, Hershey himself remarked cautiously that it was not yet clear whether his result had any fundamental significance. James Watson, on the other hand, who was also at Royaumont and who had DNA in his crosshairs, was fairly convinced.

Pauling finally came back to DNA work toward the end of November 1952. This return was spurred by an intriguing seminar at Caltech by biologist Robley Williams. Williams showed amazingly detailed electron-microscope images of a nucleic acid salt—a chemical relative of DNA. To Pauling, the images of the long, cylindrical strands, coupled with Astbury's X-ray diffraction photos, seemed to provide definitive evidence, if he needed any, for a helical molecule. Pauling also knew from the work of the organic chemist Alexander Todd that the backbone of the DNA molecule contained repeating phosphate and sugar groups.

Armed with Astbury's photos, which showed strong reflections with spacing of about 3.4 angstroms, Pauling started to perform calculations of DNA structure on November 26. Based on the density measurements of Astbury and Bell and the diameter of the strand as measured by Williams, he estimated that the length of one residue along the fiber axis was 1.12 angstroms—almost precisely one-third of the spacing in the X-ray photo (of 3.4 angstroms). This led him to a surprising conclusion: "The cylindrical molecule is formed of *three chains,* which are coiled about one another . . . each chain being a *helix.*" In other words, having convinced himself that a two-stranded helix would yield too low a density, Pauling opted for a three-stranded helical architecture. This structure became known as the triple helix.

The next problem that he had to tackle concerned the nature of the very core of the three-chain helical design—that part of the molecule closest to the axis. The question was, which of the three known components of the nucleotides (bases, sugars, or phosphate groups) formed the core? Pauling and Corey went through a mental process of elimination:

> Because of their varied nature, the purine-pyrimidine group [the bases] cannot be packed along the axis of the helix in such a way that suitable bonds can be formed between the sugar residues and the phosphate groups . . . It is also unlikely that the sugar groups constitute the core of

the molecule . . . the shape . . . is such that close packing of these groups along a helical axis is difficult, and no satisfactory way of packing them has been found . . . *We conclude that the core of the molecule is probably formed of the phosphate groups* [emphasis added].

The arrangement now looked like this: The phosphate groups were arranged about the axis of the helix, with the sugars surrounding them and the bases projecting radially outward. The three-stranded molecule was held together by hydrogen bonds between the phosphate groups of different strands.

This structure looked promising, but Pauling still saw some problems. The center of the molecule now appeared to be so jam-packed by the three chains of phosphates that it resembled the "telephone booth squash"—the competition to cram as many people as possible into a telephone booth. Pauling knew that the phosphate ion was tetrahedral in shape, with the central phosphorus atom surrounded by four oxygen atoms positioned at the vertices of a pyramid. Throughout the month of December, he, Corey, and chemist Verner Schomaker were continuously trying to squeeze, warp, and twist those tetrahedra so that they would fit better. In this process, Pauling was following the same instincts that had led him previously to triumph with the alpha-helix. He believed that if he could find a structural chemistry solution that was generally consistent with the X-ray data, all other problems would sort themselves out later. For instance, there was a question of how the model allowed for the existence of a sodium salt of DNA, since there was definitely no room for sodium ions in the core. Pauling did not have an answer, but he assumed that one would be found once the main architecture was figured out. The pace of the work was frantic. Pauling even had a small group of scientists in his lab for an informal presentation of the model on Christmas Day. By the end of the month, he thought he had gotten it to be essentially right. Pauling and Corey submitted the paper, "A Proposed Structure for the Nucleic Acids," for publication on the last day of 1952. The paper started, "The nucleic acids, as constituents of living organisms, are comparable in importance to the proteins." A few phrases with a more cautious tone followed:

> We have now formulated a promising structure for the nucleic acids . . . This is the first precisely described structure for the nucleic acids that has been suggested by any investigator. The structure accounts for some of the features of the x-ray photographs; but detailed density calculations have not yet been made, and the structure cannot be considered to have been proved to be correct.

In other words, even if some of the wrinkles still had to be ironed out, Pauling wanted to establish priority.

Contrary to the somewhat tentative spirit of the scientific paper, in his personal communications about the proposed model, Pauling expressed more confidence and was extremely upbeat. In a letter to the Scottish biochemist (and eventual Nobel laureate) Alexander Todd, dated December 19, 1952, Pauling wrote: "We have, we believe, discovered the structure of the nucleic acids. I think that it will be about a month before we send off a manuscript describing the structure, but I have practically no doubt about the correctness of the structure that we have discovered . . . The structure is really a beautiful one." In a letter sent on the same day to Henry Allen Moe, president of the Guggenheim Foundation, Pauling repeated the same sentiment: "I have now discovered, I believe, the structure of the nucleic acids themselves."

Another person with whom Pauling was corresponding regularly was his son Peter, who, as luck would have it, had arrived at Cambridge just a few months earlier to work as a research student with John Kendrew. Peter's desk was in an office with four other colleagues. In Peter's words: "To my left, near the window, was a rather noisy fellow named Francis Crick. On my right was a desk occasionally occupied by Jim Watson. Also in the room was a visiting scientist, Jerry Donohue, whom I knew well from his long association with Caltech, and Michael Bluhm, John Kendrew's research assistant." In a pre-email era, Peter, through his frequent exchange of letters with his father, became the main line of communication between Caltech and Cambridge. Consequently, as soon as Linus informed Peter of his paper on the structure of DNA, the latter asked for a copy. This was on January 13, 1953. Peter added in his letter a brief comment that spoke volumes about the pressure the British scientists were feeling: "I was told a story today. You know how children are threatened 'You had better be good or the bad ogre will come get you.' Well, for more than a year, Francis [Crick] and others have been saying to the nucleic acid people at King's 'You had better work hard or Pauling will get interested in nucleic acids.'"

Under these conditions, it was only natural that the news from Peter that Pauling had discovered the structure of DNA hit Watson and Crick like a thunderbolt. With the memory of Pauling's previous victory with the alpha-helix still fresh in the minds of everybody at Cambridge, the two young men were wondering if this was a catastrophic case of déjà vu. On January 23 Peter sent Linus another letter, this time complaining only that "I wish Jim Watson were here [Watson was on a quick visit to Milan, Italy]. It is rather dull now. Nothing to do. No interesting girls, just young affected little things only interested in sex, in an indirect manner."

The weeks between Peter's request for a copy of Pauling's paper, and the manuscript's arrival on January 28, felt like an eternity to Watson and Crick. When Peter finally brought the paper, Watson quickly pulled it out of Peter's outside coat pocket, and instantly devoured the summary and the introduction. Then, after staring at the illustrations for a few minutes, he couldn't believe his eyes. Pauling's structure, with the phosphates in the center and the bases on the outside, was strikingly similar to his and Crick's abortive model. The model was preposterously wrong!

CHAPTER 7

WHOSE DNA IS
IT ANYWAY?

Calamities are of two kinds: misfortunes to ourselves, and
good fortune to others.

— AMBROSE BIERCE

Watson did not conclude that Pauling's DNA model was wrong just because it had three strands. Pauling's nucleic acid molecule was simply not an acid at all. That is, it could not release positively charged hydrogen atoms when dissolved in water, the very definition of an acid. Instead, the hydrogen atoms were bound firmly to the phosphate groups, rendering those electrically neutral, while every elementary chemistry book (including Pauling's own book!) stated that the phosphates had to be charged negatively (the acid is highly ionized in aqueous solution). There was no way to extract those hydrogen atoms, either, since they were actually the key links holding together the three strands through hydrogen bonds.

This blunder was just too much for Watson and Crick to swallow. The world's greatest chemist constructed a completely defective model, and the model was wrong not because of some subtle biological feature but because of a major blooper in the most basic chemistry. Still incredulous, Watson rushed to Cambridge chemist Roy Markham and to the organic chemistry laboratory to check with them whether there was any doubt that DNA, as it occurs in nature, was indeed the salt of an acid. To Watson's satisfaction, they all confirmed the unthinkable: Pauling had utterly botched the chemistry.

There were only two things left to do that day. First, Crick hurried to Perutz and Kendrew to convince them that urgency was of the utmost importance. Unless he and Watson got busy with modeling immediately, he argued, it wouldn't be long before Pauling discovered his mistake and revised his model. Crick estimated that they had no more than about six weeks to come up with a correct model. Watson and Crick's second action was equally obvious to the two young men: They went to celebrate at the Eagle Pub on Bene't Street. Watson later recalled, "As the stimulation of the last several hours had made further work that day impossible, Francis and I went over to the Eagle. The moment its doors opened for the evening we were there to drink a toast to the Pauling failure."

How could this blunder have happened? Why was Pauling's model-building approach so spectacularly successful with the alpha-helix and so disastrously ineffective with the triple helix?

Anatomy of a Blunder

Let's attempt to analyze, one by one, the causes for Pauling's failing. First, there was the issue of how much time and how much thought he had actually put into solving DNA. Pauling started to think about some aspects of the DNA structure following Ronwin's paper in November 1951. However, it wasn't until November 1952, a full year later, that he commenced working earnestly on the problem. Yet by the end of December 1952, after just about a month's work, he'd already submitted his paper! Compare this to his exertion on the polypeptide structure, where he thought about the issues for about thirteen years, delaying publication several times until he was fairly confident in his model. So, even just in terms of the time invested into thinking about DNA, there is no escape from the conclusion that the DNA model was a rush job. Maurice Wilkins certainly thought so. In an interview on the history of the discovery of DNA structure, he remarked, "Pauling just didn't *try*. He can't really have spent five minutes on the problem himself." We shall return later to the question of the possible reasons for this haste and apparent lack of focus.

Second, there was a huge difference between the quality of the data on the basis of which Pauling constructed his model for proteins and his model for DNA. In the case of the alpha-helix, Pauling's collaborator, Robert Corey, had produced a vast arsenal of structural information on sizes, volumes, and angular positions for amino acids and simple peptides. For DNA, by contrast, Pauling was operating almost in a vacuum. The only X-ray photographs available to him were of poor quality and had been produced from a mixture of the A and B forms (unbeknown to him), rendering them almost useless. Worse yet, Pauling was unaware of the high water content of the preparations from which the X-ray diffraction photos were taken. By neglecting the fact that more than one-third of the material in the DNA specimens was water, Pauling obtained a wrong density, which led him to the wrong conclusion of three strands. Lastly, unlike Corey's extensive work on the building blocks of proteins, there was no equivalent effort on the bases—the subunits of the nucleotides.

Then there were the two astounding memory lapses: one about Chargaff's base ratios and one about Pauling's own self-complementarity principle. Chargaff's findings that the amount of the A base was equal to that of T, and the amount of C equal to that of G, argued for the bases somehow pairing with each other and producing two strands rather than three. Pauling claimed later that he had known about these ratios but had forgotten. Chargaff himself thought that this was *the* reason for Pauling's blunder, saying, "Pauling in *his* structural model of DNA *failed to take account* of my results. The consequence was that his model did not make sense in the light of the chemical evidence."

Pauling's second memory failure was even more astonishing. Recall that Pauling had said in 1948 that if genes consisted of two parts that were complementary to each other in structure, replication was relatively straightforward. In that case, each of the parts could serve as a mold for the production of the other part, and the complex of the two complementary parts as a whole could serve as a mold for a duplicate of itself. Clearly, this principle of self-complementarity suggested strongly a two-stranded architecture, and it was markedly at odds with a structure consisting of three strands. Yet Pauling had apparently completely forgotten this principle by the time he constructed his DNA model.

When I talked to Alex Rich and Jack Dunitz, who were Pauling's postdocs at the time, both agreed that had Pauling seen Rosalind Franklin's X-ray photograph 51 of the B form of DNA, he would have realized immediately that the molecule possessed a two-fold symmetry, pointing to a double-stranded rather than a three-chain structure. As we have seen, however, Pauling made no special effort to see Franklin's photographs.

In January 2011, I asked James Watson how surprised he was when he saw Pauling's erroneous triple-helix model. Watson laughed. "Surprised? You could not have written a fictional novel in which Linus would have made an error like this. The minute I saw that structure, I thought, 'This is wacko.'"

A close examination of the many potential causes for Pauling's calamitous model raises a series of questions at a deeper level: How can we explain the haste, the apparent lack of exertion, the forgetfulness, and the disregard for some of the basic rules of chemistry?

On the face of it, the haste is particularly puzzling if we accept Peter Pauling's testimony that there never was a "race" to solve the DNA structure. In the same entertaining account in which he noted that to his father DNA was just another interesting chemical, Peter added, "The story of the discovery of the structure of DNA has been described in the popular press as 'the race for the double helix.' This could hardly be the case. The only person who could conceivably have been racing was Jim Watson." Peter explained further that "Maurice Wilkins has never raced anyone anywhere," and that Francis Crick simply liked "to pitch his brain against difficult problems." I asked Alex Rich and Jack Dunitz about it, and neither of them thought that there was a race as far as Pauling was concerned. Why, then, did he hurry so much to publish? "Because he was always competitive," Rich suggested. This is certainly true, but it can be only part of the explanation, since Pauling had shown so much more caution and patience in the case of the alpha-helix. Ironically, his triumph with the alpha-helix had no doubt contributed to his defeat with the triple helix, since Pauling assumed, based on his success with the former, that he could reproduce the accomplishment with the latter. In this sense, this was a classical case of *inductive reasoning:* the common strategy of probabilistic guessing based on past experience—taken way too far.

Everyone engages in inductive reasoning all the time, and usually it helps us make correct decisions based on relatively little data. Suppose I ask you, for instance, to complete this sentence: "Shakespeare was a uniquely talented ___." Most people would probably answer "playwright," and they would be perfectly justified in doing so. While there is nothing illogical with completing the sentence with "cook," or "card player," chances are that the word sought for was indeed "playwright." Inductive reasoning is what allows us to use our cumulative experience to solve problems through the choice of the most likely answer. Like experienced chess players, we do not typically analyze every possible logical answer. Rather, we opt for what we think is the most probable one. This is an essential part of our cognition. Psychologist Daniel Kahneman described the process this way: "We can't live in a state of perpetual doubt, so we make up the best story possible and we live as if the story were true." However, because inductive reasoning involves probabilistic guesswork, it also means that sometimes it gets things wrong, and occasionally, it can get things *very* wrong. Pauling thought that he could take a shortcut, because past experience had shown him that all of his structural hunches turned out to be correct. In the DNA failure, the blunderer was a victim of his own previous brilliance.

Why, however, did he feel that he needed to cut corners at all? Certainly not because of Watson and Crick—he was barely aware of their endeavors—but because he did know that King's and perhaps even the Cavendish had access to superior X-ray data. He must have assumed that it would not be too long before his old rivals Bragg, Perutz, Kendrew, or perhaps Wilkins would figure out the correct structure. He decided to gamble, and he lost.

But there is very little doubt that had Pauling significantly delayed publication of his model, some researchers from Cambridge or London would have published their correct model first. Even though Pauling did not think specifically about Watson and Crick, he did know that the competition had the better hand. Therefore, taking a calculated risk may not have been altogether crazy.

On a more speculative note, Pauling's decision to rush publication may also have been related to a human cognitive bias known as the *framing effect*, which reflects a strong aversion to loss. Have you ever wondered why stores generally advertise ground beef as being "90 percent lean," rather than "10 percent fat"? People are much more likely to buy it with the former label, even though the two labels are equivalent. Similarly, people are more likely to vote for an economic agenda that promises 90 percent employment than for one that emphasizes 10 percent unemployment. Numerous studies show that the degree to which we perceive loss as devastating is higher than the degree to which we perceive an equivalent gain as gratifying. Consequently, people tend to seek risks when presented with a negative frame. Pauling may have preferred to take the risk when faced with the possibility of a probable loss.

There is also the puzzling issue of Pauling's forgetting the Chargaff rules and, more importantly, his own insights on the self-complementarity of the genetic system. I believe that the latter was a strong manifestation of the fact that even when he finally decided to work on DNA, Pauling was still not entirely convinced that this molecule truly represented the very secret of life—the mechanism of cell division and heredity. Four main clues lead me to this conclusion: (1) There is Peter's testimony that to his father DNA was just another interesting chemical and nothing more. Pauling was, after all, a chemist and not a biologist. (2) In his letter to the president of the Guggenheim Foundation announcing his "discovery" of the structure of DNA, Pauling added this, rather lukewarm, sentence:

"Biologists probably consider that the problem of the structure of nucleic acid is fully as important as the structure of proteins" (note the noncommittal flavor of the phrase "Biologists probably consider"). (3) We have the pointed question that Pauling's wife, Ava Helen, asked him after all the hoopla surrounding the publication of the Watson and Crick model had subsided: "If that was such an important problem, why didn't you work harder on it?" (4) The Pauling and Corey paper itself (on the triple helix) provides what is perhaps the most convincing piece of evidence for Pauling's lack of confidence in DNA's importance. Pauling and Corey discuss the biological implications of their model only obliquely. In the opening paragraph of their paper, they mention halfheartedly that evidence exists that the nucleic acids "are involved" in the processes of cell division and growth, and that they "participate" in the transmission of hereditary characters. Only in the last paragraph of the original manuscript do they vaguely address the topic of coding of information (but not of copying), noting, "The proposed structure accordingly permits the maximum number of nucleic acids to be constructed, providing the possibility of high specificity." I believe that this lack of conviction on Pauling's part about the crucial role of DNA was at the core of the reality that the topic of heredity—and Pauling's important pronouncements on it—apparently remained largely disconnected in his mind from the problem of the structure of DNA.

Forgetting Chargaff's rules is, in my opinion, less mysterious. First, Pauling's personal dislike for Erwin Chargaff surely contributed somewhat to his lack of attention to Chargaff's results. Second, recall that Pauling was continuously distracted during his work on DNA. Enmeshed in his attempts to complete the work on proteins and in his bitter political struggles with McCarthyism, he barely had any time left to concentrate. Actually, on March 27, 1953, just two months after Peter received the manuscript on DNA, Pauling wrote a letter to Peter in which he commented, "I am just putting the final touches on my paper on a new theory of ferromagnetism." He was already thinking of something else! This hardly could have helped.

Extensive studies by Swedish researchers showed that natural memory problems (known as benign senescent forgetting) occur much more frequently when attention is divided or has to be shifted rapidly. Therefore, Pauling's not remembering Chargaff's rules is not very surprising.

Finally, there is the truly dumbfounding question of why Pauling ignored some basic rules of chemistry in his model, such as those concerning the acidity of DNA. The world's most celebrated chemist fumbling in some elementary chemistry?

I asked molecular biologist Matthew Meselson about his thoughts on this aspect of the blunder. Meselson, Pauling's graduate student at the time, conjectured that Pauling might have considered the problem and had convinced himself that it could somehow be overcome. This would certainly be consistent with Pauling's general frame of mind throughout the entire DNA model-building episode. His thought process must have been something like this: He had a highly successful model for proteins, which consisted of a helical strand with side chains on the outside. He therefore thought that the model for DNA would be that of interwoven strands, also with side chains (the bases, in this case) on the outside. This created a packing problem along the axis, but all the rest of the characteristics, in Pauling's mind, were in some sense details to be sorted out later. Again, his previous success with the alpha-helix apparently had a blinding effect. Unfortunately, as we know only too well, the devil is often in precisely those details.

In my conversation with Jack Dunitz, he recalled that Pauling had once told him something that summarized beautifully his attitude toward scientific research:

> Jack, if you think you have a good idea, publish it! Don't be afraid to make a mistake. Mistakes do no harm in science because there are lots of smart people out there who will immediately spot a mistake and correct it. You can only make a fool of yourself and that does no harm, except to your pride. If it happens to be a good idea, however, and you don't publish it, science may suffer a loss.

Dunitz added that indeed the three-stranded structure did no harm, except to Pauling's reputation. He commented further that Pauling had made enough major contributions that we should simply forgive and forget. I must say that I fully agree with the "forgive" part, but I actually think that we should *not* forget. As I have attempted to show, there are many insights that can be gained from analyzing such blunders by brilliant individuals.

Seeing Double

The rest of the story of the discovery of the structure of DNA has been told and retold numerous times, but the recently discovered correspondence of Francis Crick does shed some new light on the frantic activity that preceded the publication of the Watson and Crick model.

Pauling's blunder served as the catalyst that convinced Bragg to allow Watson and Crick to go back to DNA modeling. Within a couple of weeks, Watson went to London, where Wilkins, also pleased with Pauling's glitch, took the liberty of showing him Franklin's famous photograph 51 of the B form of DNA (figure 14), without Franklin's knowledge. Much ink has been devoted to the question of the ethical nature of this particular act. In my humble opinion, three main parts to this story deserve attention. First, there was apparently no problem with Wilkins himself having a copy of the photo (given to him by Gosling), since Franklin was about to leave King's to work at Birkbeck College, and she had been informed by the lab director, Sir John Randall, that the results of all the DNA work belonged exclusively to King's. Second, there is little doubt (in my mind at least) that Franklin should have been consulted before her *unpublished* results were shared with members of another laboratory. Finally, there is disagreement on whether or not Watson and Crick acknowledged Franklin's contribution adequately in their paper. You can be the judge of that. They wrote, "We have also been stimulated by a knowledge of the general nature of the unpublished experimental results and ideas of Dr. M. H. F. Wilkins, Dr.

R. E. Franklin and their co-workers at King's College, London." Be that as it may, the effect the photo had on Watson was dramatic: The dark cross was the unmistakable sign of a helical structure. No wonder that, as he later described, his "mouth fell open," and his "pulse began to race."

Watson and Crick spent the following weeks trying frantically to build models in which the bases would form the rungs of the helical ladder they had in mind. The first attempts were unsuccessful. Ignoring the clue from Chargaff's ratios, Watson mistakenly thought that he should pair every base with its twin, forming rungs composed of adenine-adenine (A-A), cytosine-cytosine (C-C), guanine-guanine (G-G), and thymine-thymine (T-T). However, since the bases C and T were different in length from G and A, this created rungs of unequal lengths, which was inconsistent with the symmetric pattern exhibited in photograph 51. There was also the question of the bond between the two bases in each rung and between the rung and the "legs" of the ladder (which were supposed to be composed of sugars and phosphates). Here again Watson and Crick were heading the wrong way, but their office mate Jerry Donohue came to the rescue. As a former student of Pauling's, Donohue knew everything there was to know about hydrogen bonds. Donohue pointed out to Watson and Crick that even many textbooks had the hydrogen atoms in the wrong positions in thymine and guanine. Placing these atoms in their correct locations opened new possibilities for bonding the bases to each other. By shifting the bases in and out of other pairing possibilities (than the like-with-like), Watson suddenly realized that an A-T pair held together by two hydrogen bonds was identical to a G-C pair held similarly. The rungs became of equal length. Moreover, this pairing provided a natural explanation to Chargaff's rules. Clearly, if A always paired with T, and G with C, then the numbers of A and T molecules in any section of DNA would be equal, and similarly for G and C. Another source of valuable information became available around that time via Max Perutz: a copy of Franklin's report, written for a visit of the Medical Research Council biophysics committee to King's. From the symmetry of the crystalline

DNA described in that report, Crick concluded that the two strands of DNA were antiparallel—they ran in opposite directions.

The resulting structure was the celebrated double helix, in which the two helical strands (the backbones) were made of alternating phosphates and sugars, with the paired bases attached to the sugars and making the rungs (figure 15). At this point, Watson and Crick were so convinced of the correctness of their model that they were eager to submit a short paper to *Nature* to announce it. Even before that, according to Watson's by-now-famous description, Crick interrupted patrons' lunchtime at the Eagle to make public that he and Watson had "discovered the secret of life." Figure 16 shows the spot

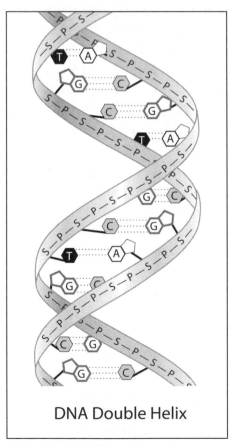

DNA Double Helix

Figure 15

Figure 16

in the Eagle where Crick made the announcement. On March 17, 1953, Crick sent a copy of the paper to Wilkins. One of the recovered documents in Crick's "lost" correspondence is a draft of the letter that was to accompany the manuscript. Part of it reads:

> *Dear Maurice,*
>
> *I enclose a draft of our letter. As it has not yet been seen by Bragg I would be grateful if you did not show it to anyone else. The object of sending it to you at this stage is to obtain your approval of two points:*
> *a) the reference number 8 to your unpublished work.*
> *b) the acknowledgement.*
> *If you would like either of these rewritten, please let us know. If we don't hear from you within a day or so we shall assume that you have no objection to their present form.*

This draft and another one addressed to one of the editors of *Nature* (which apparently was never mailed) show that Crick and Watson were at first under the impression that theirs was the only manuscript to be submitted at that time. Actually, the two groups at King's submitted papers to *Nature* as well. In a brief note to Crick probably written on the same day, Wilkins says, "Herewith almost uncorrected draft. How should we refer to your note?" This accompanied a draft of Wilkins's own manuscript. The third paper was by Rosalind Franklin and Raymond Gosling.

Once he realized the situation, Crick expressed his view that everyone should see everyone else's manuscript: "It is not reasonable for letters to be sent jointly to *Nature* without having been read by all concerned. We want to see hers [Franklin's], and I've no doubt she wishes to see ours." Wilkins agreed. In a newly found letter dated "Mon.," probably referring to Monday, March 23, he said, "We will post a copy of Rosy's thing to you tomorrow," adding, "Raymond and Rosy have your thing so everybody will have seen everybody else's."

Perhaps the most fascinating part of the new correspondence, however, is related to Pauling. First, Crick expressed his displeasure with the fact that Franklin might want to see Pauling on his forthcoming visit to England. "It is not impossible," he wrote to Wilkins, "that she might consider turning over the experimental data to Pauling. This would inevitably mean that Pauling would prove the structure and not you." To which Wilkins responded with irritation: "If Rosy wants to see Pauling, what the hell can we do about it? If we suggested it would be nicer if she didn't that would only encourage her to do so. Why is everybody so terribly interested in seeing Pauling . . . Now Raymond [Gosling] wants to see Pauling too! To hell with it all." This exchange is a perfect demonstration of the awe that Pauling continued to inspire even at one of the lowest moments in his career.

The April 25, 1953, issue of *Nature* contained three papers on the structure of DNA. First, there was the landmark paper by Watson and Crick describing the double helix structure. The paper was just a little more than one page long, but what a page that was. Watson and Crick started by acknowledging, "A structure for nucleic acid has already been proposed by Pauling and Corey. They kindly made their manuscript available to us in advance of publication." However, they immediately added, "In our opinion, this structure is unsatisfactory." They then concisely explained their "radically different structure," consisting of "two helical chains each coiled around the same axis," and, in particular, the "novel feature" of the structure, which is "the manner in which the two chains are held together by the purine and pyrimidine bases."

Watson and Crick's model immediately suggested a solution both to how the coding of genetic information is achieved and to the puzzle of how the molecule manages to copy itself. The details were presented in a second paper, published just five weeks after the first, in which Watson and Crick proposed the mechanism underlying the genetic code: "The phosphate-sugar backbone of our model is completely regular, but any sequence of the pairs of bases can fit into the structure. It follows that in a long molecule many different permutations are possible, and it therefore seems likely that *the precise sequence of the bases is the code which carries the genetical information* [emphasis added]." The message was clear: The coding of the genetic instructions that are needed to create, say, an amino acid, is contained in the specific sequence of bases in the rungs. For instance, the sequence C-G followed by G-C and then by T-A codes for forming the amino acid arginine, while G-C followed by C-G and then by T-A codes for alanine. The copying is done (precisely as anticipated abstractly by Pauling in 1948) by "unzipping" the double helix ladder at its center, producing two halves, each containing a leg and one-half of each one of the rungs. Because the sequence of bases in one chain automatically determines the sequence of the bases in the other (since the partner of T is always A, and that of G is always C), it is clear that one-half of the molecule contains all the information needed for constructing the whole molecule. For instance, if the sequence of bases along one chain of DNA is TAGCA, then the complementary sequence in the other chain must be ATCGT. This way, two new complete ladders can be generated from the original one and, hence, copying of the DNA molecule is accomplished.

In their first paper, Watson and Crick did not spell out the copying mechanism, but they remarked laconically, "It has not escaped our notice that the specific pairing we have postulated immediately suggests a possible copy mechanism for the genetic material." Crick explained later that this enigmatically economical sentence (which has been labeled "coy" by some historians of science) was, in fact, a compromise between his own desire to discuss the genetic implications in the first paper and Watson's concern that the structure might still be wrong. The statement was, therefore, a simple claim to priority. The fact that Watson did still harbor doubts about the model is well documented in his contemporary letters.

As I have noted, two other papers in *Nature* accompanied the first paper by Watson and Crick. One was by Wilkins, Alexander Stokes, and Herbert Wilson, in which they analyzed some of the X-ray crystallographic data and also presented evidence that the helical structure exists not just in isolated fibers but also in intact biological systems. In the years that followed, Wilkins and his colleagues, and also Matthew Meselson, Arthur Kornberg, and others, did much work to confirm in detail the Watson and Crick model and their conclusions.

The third paper in the April 25, 1953, issue of *Nature* was by Franklin and Gosling. It contained the famous X-ray photograph of structure B. True to Franklin's general attitude to science, the manuscript was formulated cautiously to read:

> While we do not attempt to offer a complete interpretation of the fibre-diagram of structure B, we may state the following conclusions. The structure is probably helical. The phosphate groups lie on the outside of the structural unit, on a helix of diameter about 20 Angstroms. The structural unit probably consists of two coaxial molecules which are not equally spaced along the fibre axis . . . Thus our gen-

eral ideas are not inconsistent with the model proposed by Watson and Crick in the preceding communication.

Few would disagree with the statement that Franklin's exquisite X-ray diffraction photographs provided crucial information concerning DNA's overall structure and specific dimensions. Sadly, Rosalind Franklin died from cancer in 1958 at the age of thirty-seven. It is conceivable that the disease was brought on by overexposure to those same X-rays that helped uncover the structure of DNA. Four years later, Watson, Crick, and Wilkins shared the Nobel Prize in Physiology or Medicine for discovering the molecular structure of DNA and its significance for information transfer in living material. Since the Nobel Prize is not awarded posthumously and cannot be shared by more than three people (in a given category in a given year), we shall never know what would have happened had Franklin survived until 1962.

In 2009 the famous photograph 51 became the title of a successful play by Anna Ziegler. As its title implies, the fictionalized account in the play concentrated on Rosalind Franklin and her rocky relationship with Maurice Wilkins. When asked to comment on the play, Watson remarked that the character of Maurice Wilkins "talked too much," while the actor who played Crick did not do justice to the real Crick, since the play imparted him with a "used-car-salesman" vibe.

No one likes to admit defeat, and scientists are no exception. In a letter Pauling wrote to Peter on March 27, 1953, he first "casually" noted:

> It might be good for you to get in touch with Miss Franklin, if you decide that this is a good plan, and arrange for us to see her also. If the King's College people (Miss Franklin has left King's College, and is with Bernal at Birkbeck) express an interest in having me visit their place, perhaps this could be worked in on the same day. I am not planning, however, to approach them on the matter.

Then, after another paragraph in which he described his precise travel plans, Pauling continued:

> I have received a letter from Watson and Crick, describing their structure briefly—a copy of their letter to *Nature* is enclosed. The structure seems to me to be a very interesting one, and I have no strong argument against it. I do not think that their arguments against our structure are strong ones either.

Later in the letter, Pauling recognized that the water content of the molecule could be very important: "We give an argument . . . to support the assignment of three nucleotide residues . . . However, if the specimen of reasonably dry nucleic acid contained about 30% water . . . there would be only two residues on this length." He concluded, "I think that the Wilkins photographs should settle the question definitely."

I asked Alex Rich if Pauling truly thought that he could hold on to his triple helix model, and that the double helix was uncertain. Rich's answer was fairly categorical: "Of course Pauling knew that the double helix was the correct model," he said. "All this talk about it being uncertain was just bravado." Indeed, Pauling came to Cambridge the first week in April (figure 17 shows him in 1953), and after seeing Watson and Crick's wired model and Franklin's X-ray photo, and having listened to Crick's explanation, he acknowledged graciously that the structure appeared to be correct. A couple of days later, Pauling and Bragg left for the Solvay Conference in Brussels, Belgium. At that meeting of the world's top researchers, Bragg first announced the double helix. With great style, Pauling admitted during the discussion that followed, "Although it is only two months since Professor Corey and I published our proposed structure for nucleic acid, I think that we must admit that it is probably wrong."

One could certainly argue that there was nothing particularly "brilliant" in Pauling's blunder—after all, his model was built inside out and with the wrong number of chains. But it was Pauling's

Figure 17

method, way of thinking, and previous incredible success with complex protein molecules that inspired and informed Watson and Crick. In a short article published on March 21, 1999, Watson wrote about Pauling, "Failure hovers uncomfortably close to greatness. What matters now are his perfections, not his past imperfections. I most remember Pauling from 50 years ago, when he proclaimed that no vital forces, only chemical bonds, underlie life. Without that message, Crick and I might never have succeeded."

The discovery of the structure of DNA had flung open the doors to a limitless range of research that to date culminated in April 2003 with the formal completion of the Human Genome Project—the decoding of the complete DNA of a human (although analysis of all the data will continue for many years). Along the way came many surprises. For instance, prior to the year 2000, biologists believed that the human genome contained about one hundred thousand protein-coding genes. Findings from the International Human Genome Sequencing Consortium published in October 2004 reduced the esti-

mate to fewer than twenty-five thousand—only a little more than the
gene count of the simple roundworm *C. elegans*! Cheaper and faster
genetic sequencing technology has recently helped scientists to draw
a new picture of human origins. The new view that emerges from the
genetic analysis of the tip of a girl's forty-thousand-year-old pinky
finger found in a Siberian cave, is that modern humans did not simply march out of Africa. Rather, they probably encountered and bred
with at least two other groups of ancient, now-extinct humans.

The discovery of the structure and function of DNA has also shed
light on evolution by clarifying the nature of the hereditary variations on which natural selection can operate. Pauling's proclamation
that life processes are the consequence of the laws of chemistry and
physics became verifiable through an understanding of the forces
that shape and can vary DNA patterns. (Figure 18, a picture of some
of the participants in the Pasadena Conference on the Structure of
Proteins that took place in September 1953, shows many of the major
players in the discovery of the alpha-helix and the double helix.)

1. Wilson 2. Perutz 3. Schomaker 4. Watson 5. Dunitz 6. Crick

7. Wilkins 8. Kendrew 9. Rich 10. Pauling 11. Corey 12. Astbury 13. Bragg

Figure 18

We cannot even imagine what opportunities our comprehension of DNA and our ability to modify the molecule will present in the distant future. Possibilities range from significant lengthening of the human life expectancy to the creation of new life-forms. Deciphering the DNA structure has already led to an understanding of the genetic basis of diseases, which has revolutionized the search for treatments. The genome era has heralded previously unimaginable achievements in forensic science. For instance, following the deaths of five people from anthrax-laced letters in 2001, the US Federal Bureau of Investigation decided to sequence the entire microbial genome of the strain used in the attacks (5.2 million base pairs). That effort eventually led investigators to an army lab that was the most likely source of the strain. At the same time, with the exposition of the structure of DNA and of proteins, the question of the origin of life has become even more intriguing and potentially answerable. But the inquiries have penetrated to an even more fundamental level than the purely biological: Where did the building blocks of life, those information-carrying, replicating molecules, come from? And on the physics side, going back to earlier origins yet, how did the hydrogen atom, which was so crucial to Pauling's hydrogen bond, appear in the universe? And what about the heavier elements that are so essential for life, such as carbon, oxygen, nitrogen, and phosphorus?

The Russian-born physicist George Gamow participated in the early attempts to understand how the four bases in DNA could control the synthesis of proteins from amino acids. Gamow was shown a copy of the paper by Watson and Crick on the genetic implications of their model while visiting the Radiation Laboratory at Berkeley. Excited, he started thinking about it as soon as he returned to his department at George Washington University, swiftly dispatching a letter to Watson and Crick. He started apologetically — "Dear Drs. Watson and Crick, I am a physicist, not a biologist" — but soon came to his main point: Could the relationship between the four letters corresponding to the bases in DNA, and the twenty amino acids in proteins, be solved as a problem in pure numerical cryptoanalysis? While Gamow's mathematical solutions eventually turned out to be wrong, they did help in framing the questions of biology in the language of information.

About five years earlier, Gamow was involved in solving an even more fundamental problem: the cosmic origin of hydrogen and helium. His solution was truly brilliant. It did not explain, however, the existence of all the elements heavier than helium. This formidable task was left to another astrophysicist and cosmologist: Fred Hoyle. On one hand, Hoyle concerned himself with the evolution of the universe as a whole, and on the other, with the emergence of life within it. He was at the same time one of the most distinguished and one of the most controversial scientists of the twentieth century.

CHAPTER 8

B FOR BIG BANG

The philosophy which is so important in each of us is
not a technical matter; it is our more or less dumb sense
of what life honestly and deeply means. It is only partly
got from books; it is our individual way of just seeing and
feeling the total push and pressure of the cosmos.

—WILLIAM JAMES

On March 28, 1949, at six thirty in the evening, astrophysicist Fred Hoyle gave one of his authoritative radio lectures on the BBC's *The Third Programme,* a cultural broadcast that featured such intellectuals as philosopher Bertrand Russell and playwright Samuel Beckett. At one point, as he was trying to contrast his own scenario—one of continuous creation of matter in the universe—with the opposing theory, which claimed that the universe had a distinct and definite beginning, Hoyle made what was to become a controversial statement:

> We now come to the question of applying the observational tests to earlier theories. These theories were based on the hypothesis that *all the matter in the universe was created in one big bang at a particular time in the remote past* [emphasis added]. It now turns out that in some respect or other all such theories are in conflict with the observational requirements.

This lecture marked the birth of the term "big bang," which has since been inextricably attached to the initial event from which our universe sprouted. Contrary to popular belief, Hoyle did not use the term in a derogatory manner. Rather, he was simply attempting to create a mental picture for his listeners. Ironically, a scientist who always opposed the idea behind this model coined and popularized the term big bang. The name has even survived a public referendum. In 1993 *Sky & Telescope* magazine solicited suggestions from readers for a better name—an act generally viewed as an attempt at cosmic political correctness. After three judges (including Carl Sagan, the famous astronomer and popularizer of science) sifted through the 13,099 entries, however, they found no worthy replacement. The title of this chapter ("*B* for Big Bang") was fashioned after the title of a British television science-fiction drama, *A for Andromeda*, written by Hoyle and TV producer John Elliot. The seven-part series aired in 1961, and it featured actress Julie Christie in her first major role.

Fred Hoyle was born on June 24, 1915, in Gilstead, a village near the town of Bingley in West Yorkshire, England. His father was a wool and textiles merchant who was drafted into the Machine Gun Corps and dispatched to France during World War I. His mother studied music, and for a while played the piano in a local cinema, to accompany silent films. Fred Hoyle, who originally planned to be a chemist, studied mathematics at Cambridge, and he demonstrated such talent and accomplishments that he was elected fellow of St. John's College in Cambridge in 1939. In 1958 he earned the prestigious Chair of Plumian Professor of Astronomy and Experimental Philosophy at Cambridge. Incidentally, George Darwin, Charles Darwin's son, had held this chair between 1883 and 1912.

Signs of Hoyle's relish for independence and sometimes dissension were apparent from an early age. He later recalled: "Between the ages of five and nine, I was almost perpetually at war with the education system . . . As soon as I learned from my mother that there was a place called school that I must attend willy nilly—a place where you were obliged to think about matters prescribed by a 'teacher,' not about matters decided by yourself—I was appalled." His disdain for convention continued into his university years. In 1939 he decided to forgo a PhD degree for the "earthy motive," in his words, of having to pay less income tax!

Not surprisingly, this curiosity-driven independent thinker matured to become a brilliant scientist. In terms of contributions to astrophysics and cosmology, Hoyle was probably the leading figure for at least a quarter century. At the same time, he never shied away from controversy. "To achieve anything really worthwhile in research," he once wrote, "it is necessary to go against the opinions of one's fellows. To do so successfully, not merely becoming a crackpot, requires fine judgment, especially on long-term issues that cannot be settled quickly." We shall soon discover that Hoyle followed his own advice to a fault.

Even without World War II, 1939 was a critical year for Hoyle. It so happened that one after another, two of his research supervisors left Cambridge for appointments elsewhere. His third advisor was the great physicist Paul Dirac, one of the founders of quantum mechanics—the revolutionary new view of the subatomic microworld. Following the wealth of novel ideas of the 1920s, science of the late 1930s looked dull by comparison. Hoyle later wrote that Dirac told him one day in 1939, "In 1926 it was possible for people who were not very good to solve important problems, but now people who *are* very good cannot find important problems to solve." Hoyle took this warning to heart and shifted his focus from pure, theoretical nuclear physics to the stars.

Out of Hoyle's numerous accomplishments, I want to concentrate here on only a few of his contributions to one particular topic: nuclear astrophysics. Hoyle's work in this area has become one of the main pillars on which our modern understanding of stars and their evolution rests. Along the way, he solved the puzzle of how the atoms of carbon, the anchor of complexity and life as we know it, formed in the universe. To fully appreciate the significance of Hoyle's achievement, however, we first need to understand the background against which he produced his masterwork.

Prologue to the History of Matter

On one of the walls of almost every science classroom in the United States, you can find a chart of the periodic table of the elements (figure 19). Just as our language consists of words constructed from the letters of the alphabet, all ordinary matter in the cosmos is composed of these elements. Elements are those substances that cannot be further broken down or modified by simple chemical means. Dmitry Mendeleyev, a Russian chemist, is generally credited with having noticed (in the mid–nineteenth century) the periodic regularities that are the basis of the periodic table, and with having the foresight to predict the characteristics of elements that had yet to be discovered to complete the table. In many ways, the periodic table is a symbolic representation of the progress achieved since Empedocles' and Plato's famous fire, air, water, and earth as the basic constituents of matter.

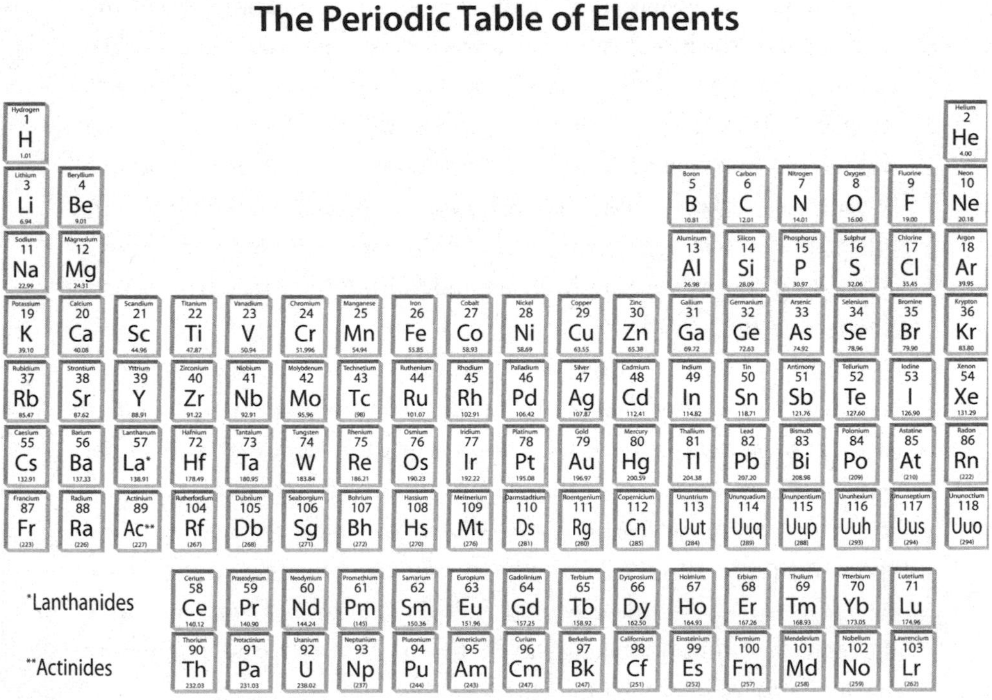

The Periodic Table of Elements

Figure 19

As an amazing aside, the smallest reproduction of the periodic table was engraved in 2011 onto a human hair belonging to chemist Martyn Poliakoff of the University of Nottingham in the United Kingdom. The engraving was done at the university's nanotechnology center. (The hair was then returned to Poliakoff as a birthday gift.)

The periodic table currently consists of 118 elements (the latest, ununoctium, was identified in 2002), of which 94 occur naturally on Earth. If you think about it for a moment, this is a fairly large number of primary building blocks, and consequently, it was only a matter of time before someone would ask, Where did all of these chemical elements come from? Or: Could these rather complex entities have simpler origins?

Someone actually did pose these questions even before the publication of the periodic table. In two papers published in 1815 and 1816, the English chemist William Prout hypothesized that the atoms of all the elements were in fact condensations of different numbers of hydrogen atoms. Astrophysicist Arthur Eddington combined the general idea of Prout's hypothesis with some experimental results on nuclei by physicist Francis Aston to formulate his own conjecture. Eddington proposed in 1920 that four hydrogen atoms could somehow combine to form a helium atom. The small difference between the total mass of the four hydrogen atoms and the mass of one helium atom was supposed to be released in the form of energy, through Einstein's celebrated equivalence between mass and energy, $E = mc^2$ ("E" denotes energy, "m" is mass, and "c" is the speed of light). Eddington estimated that in this way the Sun could shine for billions of years by converting only a few percent of its mass from hydrogen into helium. Less widely known is the fact that the French physicist Jean-Baptiste Perrin expressed very similar ideas around the same time. A few years later, Eddington further speculated that stars such as the Sun could provide natural "laboratories" in which nuclear reactions could somehow transform one element into another. When some physicists at the Cavendish Laboratory objected that the Sun's internal temperature was insufficient to make two protons overcome their mutual electrostatic repulsion,

Eddington is famously said to have advised them to "go and find a hotter place." The hypothesis of Eddington and Perrin marked the birth of the idea of stellar *nucleosynthesis* in astrophysics: the notion that at least some elements could be synthesized in the hot interiors of stars. As you might have guessed from the above, Eddington was one of the strongest champions of Einstein's theory of relativity (especially general relativity). On one occasion, physicist Ludwik Silberstein approached Eddington and told him that people believed that only three scientists in the entire world understood general relativity, Eddington being one of them. When Eddington didn't answer for a while, Silberstein encouraged him, "Don't be so modest," to which Eddington replied, "On the contrary. I'm just wondering who the third might be." Figure 20 shows Eddington with Einstein at Cambridge.

To continue the story of the formation of the elements, we need to remind ourselves of some of the very basic properties of atoms. Here is an extraordinarily brief refresher. All ordinary matter is composed

Figure 20

of atoms, and all atoms have at their centers tiny nuclei (the atomic radius is more than 10,000 times the nuclear radius), around which electrons move in orbital clouds. The constituents of the nucleus are protons and neutrons, which are very similar in mass (a neutron is slightly heavier than a proton), each of them being about 1,840 times more massive than an electron. While neutrons bound in stable nuclei are stable, a free neutron is unstable—it decays with a mean lifetime of about fifteen minutes into a proton, an electron, and a virtually invisible, very light, electrically neutral particle called an antineutrino. Neutrons in unstable nuclei can decay in the same fashion.

The simplest and lightest atom that exists is the hydrogen atom. It consists of a nucleus that contains only one proton. A single electron revolves around this proton in orbits the probability for which can be calculated using quantum mechanics. Hydrogen is also the most abundant element in the universe, constituting about 74 percent of all the ordinary (known as *baryonic*) matter. Baryonic matter is the stuff that makes up stars, planets, and human beings. Moving from left to right along rows in the periodic table (figure 19), in each step, the number of protons in the nucleus increases by one, as does the number of orbiting electrons. Since the number of protons is equal to the number of electrons (and they carry opposite electric charges that are equal in magnitude), atoms are electrically neutral in their unperturbed state.

The element following hydrogen in the periodic table is helium, which has two protons in its nucleus. In addition, the helium nucleus also contains two neutrons (which carry no net electric charge). Helium is the second most abundant element, making up about 24 percent of the cosmic ordinary matter. Atoms of the same chemical element have the same number of protons, and this number is called the *atomic number* of that element. Hydrogen has the atomic number 1, helium is 2, iron is 26, uranium is 92. The total number of protons and neutrons in the nucleus is called the *atomic mass*. Hydrogen has the atomic mass of 1; helium, 4; carbon (which has six protons and six neutrons), 12. Nuclei of the same chemical element can have different numbers of neutrons, and those are called *isotopes* of that element. For instance, neon (which has ten protons), can have isotopes with ten, eleven, or twelve neutrons in the nucleus. The common notation for these different isotopes is ^{20}Ne, ^{21}Ne, and ^{22}Ne. Similarly, hydrogen (one proton, or ^{1}H) also has in nature an isotope usually called deuterium (one proton and one neutron in the nucleus, or ^{2}H), and an isotope called tritium (one proton and two neutrons, or ^{3}H).

Returning now to the central problem of the synthesis of the different elements, the physicists of the first half of the twentieth century were faced with a series of questions related to the periodic table. First and foremost: How were all of these elements formed? But also: Why are some elements, such as gold and uranium, extremely rare (hence, their high price!), while others, such as iron or oxygen, are much more common? (Oxygen is about a hundred million times more common than gold.) Or: Why are stars composed mostly of hydrogen and helium?

Since their inception, ideas about the process of the formation of the elements have been linked intimately to those on the enormous *energy sources* of stars. Recall that Helmholtz and Kelvin proposed that the Sun's power comes from slow contraction and the associated release of gravitational energy. However, as Kelvin had clearly demonstrated, this reservoir could provide for the Sun's radiation for only a limited time: no more than a few tens of millions of years. This limit was disturbingly at odds with geological and astrophysical evidence that was pointing with increasing accuracy to ages of billions of years for both the Earth and the Sun. Eddington was fully aware of this glaring discrepancy. In his address to the British Association for the Advancement of Science meeting in Cardiff, Wales, on August 24, 1920, he made the following prescient statement:

> Only the inertia of tradition keeps the contraction hypothesis alive—or rather, not alive, but an unburied corpse. But if we decide to inter the corpse, let us frankly recognize the position in which we are left. A star is drawing on some vast reservoir of energy by means unknown to us. *This*

reservoir can scarcely be other than the subatomic energy
which, it is known, exists abundantly in all matter [empha-
sis added].

Despite his enthusiasm for the idea that stars could derive their
power from four hydrogen nuclei fusing together to assemble a
helium nucleus, Eddington had no specific mechanism for this process to actually take place. In particular, the problem of the mutual
electrostatic repulsion, mentioned above, had to be solved. Here is
the obstacle: Two protons (the nuclei of hydrogen atoms) repel each
other electrostatically because both have positive electric charges.
This Coulomb force (named after the French physicist Charles-
Augustin de Coulomb) has a long range, and it is therefore the dominant force between protons at distances larger than the size of the
atomic nucleus. Within the nucleus, however, the strong, attractive nuclear force takes over, and it can overcome the electric repulsion. Consequently, in order for protons in the cores of stars to
fuse together as envisioned by Eddington, they need to have sufficiently high kinetic energies in their random motions to overcome
the "Coulomb barrier" and allow them to interact via the attractive
nuclear force. The apparent snag in Eddington's hypothesis was that
the temperature calculated for the center of the Sun was not high
enough to impart protons with the necessary energy. In classical
physics, this would have been a death sentence for this scenario; particles with insufficient energy to overcome such a barrier just cannot
make it. Fortunately, quantum mechanics—the theory that describes
the behavior of subatomic particles and light—came to the rescue.
In quantum mechanics, particles can behave like waves, and all processes are inherently probabilistic. Waves are not precisely localized
like particles but are spread out. In the same way that some parts
of an ocean wave crashing against a seawall can splash to the other
side, there is a certain (albeit small) probability that even protons
with insufficient energy to overcome their Coulomb barrier would
still interact. Using this quantum mechanical effect of "tunneling"
through barriers, physicist George Gamow and, independently, the

two teams of Robert Atkinson and Fritz Houtermans, and Edward Condon and Ronald Gurney, demonstrated in the late 1920s that under the conditions prevailing in stellar interiors, protons could indeed fuse.

Physicists Carl Friedrich von Weizsäcker in Germany, and Hans Bethe and Charles Critchfield in the United States, were the first to elaborate the precise nuclear reactions network through which four hydrogen nuclei coalesce to form a helium nucleus. In a remarkable paper published in 1939, Bethe discussed two possible energy-producing paths in which hydrogen could convert into helium. In one, known as the proton-proton (p-p) chain, two protons first combine to form deuterium—the isotope of hydrogen with one proton and one neutron in its nucleus—followed by the capture of an additional proton that transforms the deuterium into an isotope of helium. The second mechanism, known as the carbon-nitrogen (CN) cycle, was a cyclic reaction in which carbon and nitrogen nuclei acted only as catalysts. The net result was still the fusion of four protons to form one helium nucleus, accompanied by the release of energy. While Bethe thought originally that the CN cycle was the main mode by which our own Sun produces its energy, experiments at the Kellogg Radiation Laboratory at Caltech showed later that it was the p-p chain that mostly powered the Sun, with the CN cycle starting to dominate energy production only in more massive stars.

You have probably noticed that, as its name implies, the CN cycle requires the presence of carbon and nitrogen atoms as catalytic agents. Yet Bethe's theory fell short of demonstrating how carbon or nitrogen formed in the universe in the first place. Bethe did consider the possibility that carbon could be synthesized from the fusion of three helium nuclei together. (A helium nucleus contains two protons, and a carbon nucleus, six.) However, after completing his calculations, he asserted, "There is no way in which nuclei heavier than helium can be produced permanently in the interior of stars under the present conditions"—that is, with densities and temperatures such as those encountered in most Sun-like stars. Bethe concluded: "We must assume that the heavier elements [than helium] were built up *before* the stars reached their present state of temperature and density."

Bethe's pronouncement created a serious conundrum, since astronomers and Earth scientists were concluding at the time that the different chemical elements had to have, by and large, a common origin. In particular, the fact that atoms such as carbon, nitrogen, oxygen, and iron appeared to have approximately the same relative abundances all across the Milky Way galaxy clearly hinted at the existence of some universal formation process. Consequently, if they were to accept Bethe's adjudication, physicists had to come up with some common synthesis that could have operated before present-day stars reached their equilibrium.

Just as the theory seemed to be heading toward a paralyzing impasse, the versatile George Gamow (usually known to his colleagues as Geo) and his PhD student Ralph Alpher advanced what appeared to be a brilliant idea: Perhaps the elements could have been formed in the initial, extremely hot and dense state of the universe known as the big bang. The concept itself was genius in its clarity. In the primeval, dense fireball, Gamow and Alpher argued, matter consisted of a highly compressed neutron gas. They referred to this primordial substance as *ylem* (from the ancient Greek *yle* and the medieval Latin *hylem*, both meaning "matter"). As these neutrons started decaying into protons and electrons, all the heavier nuclei could, in principle, be produced by the successive capture of one neutron at a time from the remaining sea of neutrons (and the subsequent decay of those neutrons into protons, electrons, and antineutrinos). Atoms were supposed to march in this way up the periodic table, climbing one step with each consecutive neutron capture. The entire process was assumed to be controlled by the probability for particular nuclei to capture another neutron, and also by the expansion of the universe (discovered in the late 1920s, as we'll discuss in the next chapter). The cosmic expansion determined the overall decrease of the density of matter with time, and thereby the slowing down of the nuclear reaction rates. Alpher carried out most of the computations, and the results were published in the April 1,

1948, issue of the *Physical Review*. (April Fool's Day was Gamow's favorite publication date.) The always-whimsical Geo noticed that if he could add Hans Bethe (who had nothing to do with the calculations) as a coauthor of the paper, the three names—Alpher, Bethe, Gamow—would correspond to the first three letters of the Greek alphabet: alpha, beta, gamma. Bethe agreed for his name to be included, and the paper is often referred to as the "alphabetical article." Later in the same year, Alpher collaborated with physicist Robert Herman to predict the temperature of the residual radiation from the big bang, known today as the *cosmic microwave background*. (Geo, who never abandoned his lifelong interest in punning, joked in his book *The Creation of the Universe* that Robert Herman "stubbornly refuses to change his name to Delter"—to correspond to delta, the fourth letter in the Greek alphabet.)

As ingenious as the scheme of Alpher and Gamow was, it soon became clear that while nucleosynthesis in a hot big bang could indeed account for the relative abundances of the isotopes of hydrogen and helium (and some lithium and traces of beryllium and boron), it ran into insuperable problems producing the heavier elements. The challenge is easy to understand using a simple mechanical metaphor: It is very difficult to climb a ladder when some of the rungs are missing. *In nature, there are no stable isotopes with an atomic mass of 5 or 8.* That is, helium has only stable isotopes with atomic masses of 3 and 4; lithium has stable isotopes with atomic masses of 6 and 7; beryllium's only truly stable isotope has an atomic mass of 9 (atomic mass 10 is unstable but long lived), and so on. Atomic masses of 5 and 8 are missing. Consequently, helium (atomic mass, 4) cannot capture another neutron to produce a nucleus that would be sufficiently long lived to continue the neutron-capture scheme. Lithium has a similar difficulty because of the gap at atomic mass 8. The mass gaps therefore frustrated further progress along the Gamow and Alpher approach. Even the great physicist Enrico Fermi, who examined the problem in some detail with a colleague, concluded with disappointment that synthesis in the big bang was "incapable of explaining the way in which the elements have been formed."

Fermi's conclusion that carbon and heavier elements could not be produced in the big bang combined with Bethe's assertion that these elements could not be produced in stars such as the Sun created a perplexing mystery: Where and how were the heavy elements synthesized? This was the point at which Fred Hoyle entered the picture.

And God Said: "Let There Be Hoyle"

In the late fall of 1944, Hoyle's wartime activities in naval radar took him to the United States, where he used the opportunity to meet with one of the most influential astronomers of the time, Walter Baade, at the Mount Wilson Observatory in California. At the time, this observatory contained the largest telescope in the world. From Baade, Hoyle learned how enormously dense and hot the cores of massive stars can become during the late stages in their lives. Examining those extreme conditions, he realized that at temperatures approaching a billion degrees, protons and helium nuclei could easily penetrate the Coulomb barriers of other nuclei, resulting in such a high frequency of nuclear reactions and back-and-forth exchanges that the entire ensemble of particles could reach a state known as *statistical equilibrium*.

In nuclear statistical equilibrium, while nuclear reactions continue to occur, each reaction and its inverse occur at the same rate, so that there is no further overall net change in the abundances of the elements. Consequently, Hoyle argued, he could use the powerful methods of the branch of physics known as statistical mechanics to estimate the relative abundances of the various chemical elements. To actually perform the calculations, however, he needed to know the masses of all the nuclei involved, and that information was not available to him during the war years. Hoyle had to wait until the spring of 1945 to obtain a table of the masses from nuclear physicist Otto Frisch. The result of the ensuing calculation was an epoch-making paper published in 1946, in which Hoyle delineated the framework of a theory for the formation of the elements from carbon and higher in stellar interiors. The idea was mind boggling: Carbon, oxygen, and iron did not always exist (in the sense of having been formed in the big bang). Rather, these atoms, all of which are essential for life, were forged inside the nuclear furnaces of stars. Think about this for a moment: The individual atoms that currently form the two strands of our DNA may have originated billions of years ago in the cores of different stars. Our entire solar system was assembled some 4.5 billion years ago from a mixture of ingredients cooked inside previous generations of stars. Astronomer Margaret Burbidge, who was to collaborate with Hoyle a decade later, gave a wonderful description of her experience of listening to Hoyle at a meeting of the Royal Astronomical Society in 1946: "I sat in the RAS auditorium in wonder, experiencing that marvelous feeling of the lifting of a veil of ignorance as a bright light illuminates a great discovery."

Scrutinizing the consequences of his embryonic theory, Hoyle was gratified to discover a marked peak in the abundances of the elements neighboring iron in the periodic table, just as the observations seemed to indicate. This consistency of the "iron peak," as it came to be known, indicated to Hoyle that he was doing something right. However, those missing rungs in the ladder—the absence of stable nuclei at atomic masses 5 and 8—continued to beleaguer any attempt to construct a detailed (as opposed to a skeletal) network of nuclear reactions that would produce all the elements.

To circumvent the mass-gap problem, Hoyle decided in 1949 to reexamine the possibility (previously aborted by Bethe) of fusing three helium nuclei to create the carbon nucleus, and he assigned this problem to one of his PhD students. Since helium nuclei are also known as alpha particles, the reaction is usually referred to as the *triple alpha* (3α) process. As it so happened, that particular student decided to ditch his PhD work before completing it (he was Hoyle's only student to ever do so), but he failed to cancel his formal registration. The rules of academic etiquette set for such cases by the University of Cambridge were clear: Hoyle was not allowed even to touch the problem until either the student or an independent researcher published the results. Eventually, two astrophysicists published results, although the work of one of them went almost entirely unnoticed.

The Estonian-Irish astronomer Ernst Öpik proposed in 1951 that in the contracting cores of evolved stars (the stars themselves expand to become red giants), the temperature could reach a few hundred million degrees. At these temperatures, Öpik argued, most of the helium would fuse into carbon. Since, however, Öpik's paper was published in the relatively little-known *Proceedings of the Royal Irish Academy,* not many astrophysicists knew about it.

Astrophysicist Edwin Salpeter, then at the beginning of his career at Cornell University, didn't know about it either. In the summer of 1951, Salpeter was invited to visit the Kellogg Radiation Laboratory at Caltech, where the ebullient nuclear astrophysicist Willy Fowler and his group were becoming deeply involved in the study of nuclear reactions thought to be important for astrophysics. Starting with the same idea as Öpik, Salpeter examined the triple alpha process in the hot inferno at the centers of red giants—precisely the problem abandoned by Hoyle's graduate student. Salpeter immediately recognized that three helium nuclei could hardly be expected to collide simultaneously. It was more likely that two of them might stick together long enough to be struck by a third. Salpeter soon found that carbon could perhaps be produced via a low-probability, two-step process. In the first step, two alpha particles could combine to form a highly unstable isotope of beryllium (^{8}Be), and in the second, the beryllium could capture a third alpha particle to form carbon. But there was still a serious problem. Experiments had shown that this particular isotope of beryllium disintegrates back into two alpha particles, with a fleeting mean lifetime of only about 10^{-16} seconds (0.00 ... 1 at the sixteenth decimal place). The question was whether at a temperature of over one hundred million Kelvin, the reaction rate could become so high that some of these ephemeral beryllium nuclei could fuse with the third helium nucleus before falling apart.

When he read Salpeter's paper, Hoyle's first reaction was anger with himself for having let such an important calculation slip through his fingers because of the mishap with the graduate student.

Upon a closer examination of the entire nuclear reactions network, however, Hoyle estimated that under Salpeter's assumptions, all the carbon would be transformed into oxygen essentially as fast as it was produced, by fusing with yet another helium nucleus. Some thirty years later, he described this important realization: "Bad luck for poor old Ed, I thought to myself." (Ed Salpeter was, in fact, nine years younger than Hoyle.) But did this spell disaster for the entire scheme? These were precisely the types of situations in which Hoyle revealed his incredible physical intuition and the clarity of his thought. He started with the obvious: "There has to be some way of synthesizing ^{12}C." After all, not only was carbon relatively abundant in the universe, but carbon was also crucial for life. After evaluating all the potential reactions in his head, Hoyle concluded: "Nothing was better than 3α." So how could the carbon be prevented from slipping away into oxygen? In Hoyle's mind, there was only one way: "3α *had to go a lot faster than it had been calculated to do* [emphasis added]." In other words, beryllium and helium had to be able to fuse together so easily and so quickly that carbon would be produced at a much faster rate than it was destroyed. But what could substantially speed up the rate of carbon synthesis? Nuclear physicists knew of one thing: a "resonant state" in the carbon nucleus. Resonant states are values of the energy at which the probability for a reaction reaches a peak. Hoyle realized that if the carbon nucleus happened to have an energy level that perfectly matched the energy equivalent of the combined masses of the beryllium nucleus and an alpha particle (plus their kinetic energy of motion), then the rate for the fusion of beryllium with an alpha would increase significantly. That is, the probability for the unstable beryllium nucleus to absorb another helium nucleus (alpha particle) to form carbon would be enhanced greatly. But Hoyle did more than merely point out that a resonance would help. He calculated *precisely* the necessary energy level in the carbon nucleus to obtain the desired effect. Nuclear physicists measure energies in nuclei in units called MeV (an MeV is one million electron volts). Hoyle calculated that for carbon production to match the observed cosmic abundance, a resonant state in ^{12}C was needed, at about 7.68 MeV above the lowest energy level (the ground state) of the carbon nucleus. Furthermore, using the known symmetry of the ^{8}Be and ^{4}He nuclei, he predicted the quantum mechanical properties of this resonant state.

This was all very impressive, except for one "small" problem: No such state was known to exist! The mere idea that Hoyle would be using general astrophysical evidence to make an extremely precise prediction in nuclear physics (much more precise, in fact, than could be calculated based on nuclear physics) was nothing short of preposterous, but Hoyle never lacked chutzpah.

The time was January 1953, and Hoyle was spending a sabbatical of a few months at Caltech. Armed with his new prediction for an unknown energy level of the carbon nucleus, Hoyle went straight into Willy Fowler's office at Kellogg Laboratory to see whether Fowler and his group could run experiments to verify the prediction. What happened at that meeting has become legendary. Fowler recalled, "Here was this funny little man who thought that we should stop all this important work that we were doing otherwise and look for this state, and we kind of gave him the brushoff. Get away from us, young fellow, you bother us."

Hoyle himself remembered the meeting in a more positive light:

> To my surprise, Willy didn't laugh when I explained the difficulty. I cannot remember whether he called in a Kellogg mob [the nuclear physics group that included, among others, Ward Whaling, William Wenzel, Noel Dunbar, Charles Barnes, and Ralph Pixley] there and then, or whether it was a few hours or a day or two later . . . It was then that general consensus decided a new experiment should be done.

In an interview in 2001, neither Ward Whaling nor Noel Dunbar remembered any specific details of this meeting, but Charles Barnes recalled that Willy's rather small office was packed and that "as Fred presented his ideas, it was clear that the audience was visibly skeptical. Even Willy seemed to be somewhat skeptical." Whatever precisely happened at that meeting, the net result was that the "Kellogg mob" did decide to perform the experiment, and Ward Whaling and his colleagues were identified as the group that had the best experimental setup to perform the necessary measurements.

Whaling, Dunbar, and their collaborators decided to tackle the problem by bombarding nitrogen (^{14}N) nuclei with deuterium (^{2}H). This nuclear reaction produces carbon (^{12}C) nuclei and alpha particles (^{4}He). By examining carefully the energy of the outstreaming alpha particles (and remembering that the total energy is conserved), they could detect not only particles coming out with high energy (therefore leaving the carbon in its low-energy ground state) but also particles emerging with lower energy, indicating that some energy was left in the carbon nucleus. The results were clear. Within a couple of weeks, the experimental group found a resonance in carbon at 7.68 MeV (with a possible error of 0.03 MeV)—in incredible agreement with Hoyle's prediction! In their just-over-one-page paper describing the results, the nuclear physicists started by noting: "Hoyle explains the original formation of elements heavier than helium by this process" (fusion of beryllium with helium). They finished with an acknowledgement: "We are indebted to Professor Hoyle for pointing out to us the astrophysical significance of this level."

Despite his amazingly successful prediction, Hoyle realized that this was not the time to rest on his laurels. For carbon to survive, the nuclei had to obey yet another important requirement: Carbon had to be unable to rapidly capture a fourth alpha particle that would have transformed it all into oxygen. In other words, one had to be sure that there is no resonant state in the oxygen nucleus to enhance the carbon-plus-alpha reaction rate. To complete his triumph with the theory of carbon production, Hoyle showed that such a resonant reaction indeed does not occur—the energy of the respective level in oxygen is lower by about 1 percent from the value that would have made it resonant.

One might have thought that with such a coup under his belt, Hoyle would immediately rush to announce it to the world. In reality, more than a half year passed from the confirmation of his prediction until Hoyle reported on it briefly at a meeting of the American Physical Society in Albuquerque. Even in subsequent years, Hoyle never made a big deal of his remarkable achievement. In 1986 he commented:

> In a sense this was but a minor detail. But because it was seen by physicists as an unusual and successful prediction it had a disproportionate effect in converting them from the currently held view that the elements were all synthesized in the early moments of a hot universe to the more mundane view that the elements are synthesized in stars.

Others did not think that this was "but a minor detail." When the boisterous George Gamow came to summarize his own views on Hoyle's role in the theory of the formation of the elements, he did so by a witty account that he entitled "New Genesis":

> In the beginning God created radiation and ylem. And ylem was without shape or number, and the nucleons were rushing madly over the face of the deep. And God said: "Let there be mass two." And there was mass two. And God saw deuterium, and it was good. And God said: "Let there be mass three." And God saw tritium and tralphium [Gamow's nickname for the isotope of helium ^{3}He] and they were good. And God continued to call number after number until He came to the transuranium elements. But when He looked back on his work He found that it was not good. In the excitement of counting, He missed calling for mass five and so, naturally, no heavier elements could have been formed. God was very much disappointed, and wanted first to contract the universe again, and to start all over from the beginning. But it would be much too simple. Thus being almighty, God decided to correct His mistake in a most impossible way.

And God said: "Let there be Hoyle." And there was Hoyle. And God looked at Hoyle . . . and told him to make heavy elements in any way he pleased. And Hoyle decided to make heavy elements in stars, and to spread them around by supernovae explosions. But in doing so he had to obtain the same abundance curve which would have resulted from nucleosynthesis in ylem, if God would not have forgotten to call for mass five. And so, with the help of God, Hoyle made heavy elements in this way, but it was so complicated that nowadays neither Hoyle, nor God, nor anybody else can figure out exactly how it was done.

Note, by the way, that according to this "New Genesis," even God made a blunder!

The Royal Swedish Academy of Sciences also did not think that Hoyle's prediction was merely a minor detail. In 1997 it decided to give the prestigious Craford Prize (awarded in disciplines chosen to complement those for which the Nobel prizes are given) to Hoyle and Salpeter "for their pioneering contribution to the study of nuclear processes in stars and stellar evolution." In their announcement of the prize, the academy noted: "Perhaps his [Hoyle's] most important single contribution within this field was a paper where he demonstrated that the existence of carbon in Nature implied the existence of a certain excited state in the carbon nuclei above the ground state. This prediction was later verified experimentally."

Hoyle followed up on his prediction for the carbon level with a paper that established the foundation for the theory of nucleosynthesis in stars: the concept that most chemical elements and their isotopes were synthesized from hydrogen and helium by nuclear reactions within massive stars. In this paper, published in 1954, Hoyle explained how the abundances of heavy elements today are the direct products of *stellar evolution*. Stars spend their lives in a continuous battle against gravity. In the absence of an opposing force, gravity would cause any star to collapse to its center. By "igniting" nuclear reactions in their cores, stars create extremely high temperatures, and the associated high pressures support the stars against their own weight. Hoyle described how after each central nuclear fuel is consumed (first, hydrogen is fused into helium, then helium into carbon, then carbon into oxygen, and so on), gravitational contraction causes the temperature in the core to increase until the "ignition" of the next nuclear reaction. This way, Hoyle reasoned, new elements are synthesized, all the way up to iron, in each successive core-burning episode. Since each burning core is smaller than the preceding one, the star develops an onionskin-like structure, in which each layer is composed of the main product—"ashes," if you like, of the preceding nuclear reaction (figure 21). Since iron is the most stable nucleus, once an iron core forms, no more nuclear energy is available from fusion of nuclei into heavier ones. Without a source of internal heat to combat gravity, the stellar core collapses, triggering a dramatic explosion. These so-called supernova explosions powerfully eject all the forged elements into interstellar space, where

The Internal Structure of a Pre-Supernova Star

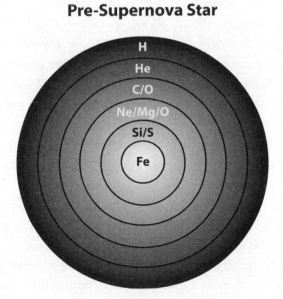

Figure 21

they enrich the gas from which later generations of stars and planets form. The temperatures attained during the explosions are so high that elements heavier than iron are formed by neutrons bombarding the stellar material. Hoyle's scenario remains to this day the broad picture depicting the evolution of stars. Surprisingly, this key paper in the development of the theory of stellar nucleosynthesis received relatively little attention at the time, perhaps because it was published in a new astrophysical journal that was relatively unknown to the nuclear physics community.

Willy Fowler was also impressed with Hoyle's prediction of the resonant level in carbon. In fact, he spent his following sabbatical in Cambridge to work with Hoyle. The collaboration between the two men and between them and the husband-and-wife team of astronomers Geoffrey and Margaret Burbidge led to one of the best-known works in astrophysics. The 1957 landmark paper by Burbidge, Burbidge, Fowler, and Hoyle—often referred to as B^2FH—gave a comprehensive theory for the synthesis of all the elements heavier than boron in stars. In a way, when Joni Mitchell sang "We are stardust," she was simply giving a concise, lyrical summary of Hoyle's 1954 paper and that of B^2FH. The four researchers used extensive astronomical data on the abundances of heavy elements in stars and meteorites and combined those with crucial nuclear data from experiments and from the hydrogen bomb test on the Eniwetok Atoll in the Pacific on November 1, 1952, to support their theoretical calculations. They described no fewer than eight nuclear processes that synthesize the elements in stars and identified the different astrophysical environments in which these processes take place. B^2FH pointed out correctly that the observational evidence that "there are real differences in chemical composition between stars" provides a strong argument in favor of a stellar synthesis theory, rather than having all the elements synthesized in the big bang.

This was a genuine tour de force. The massive, 108-page paper started with a romantic touch: two contradictory quotes from Shakespeare about the question of whether the stars govern humanity's fate. The first, from *King Lear,* reads: "It is the stars, the stars above us, govern our conditions." This is followed by the words "but perhaps" preceding the second quote, from *Julius Caesar:* "The fault, dear Brutus, is not in our stars, but in ourselves." The paper ended with a call to observers to make every possible effort to determine the relative abundances of different isotopes in stars, since those could truly be used to test the different nuclear reaction schemes. Figure 22 shows a picture taken at the Institute of Theoretical Astronomy in Cambridge in 1967. Fred Hoyle is in the middle of the second row, with Margaret Burbidge to his left. Willy Fowler is in the middle of the front row, with Geoff Burbidge to his right.

There was one thing that the B²FH paper did not achieve. No

I.O.T.A. Group 1967

M. Gribbin, J.R. Gribbin, M.Woods, R.Harding, R.C.Fisher, N.Butler, C.Nicholls.
P.Solomon, R.V.Wagoner, F.Westwater, F.Hoyle, E.M.Burbidge, D.D.Clayton, P.H.Fowler,
S.A.Colgate, G.R.Burbidge, W.A.Fowler, J.N.Islam, J.Faulkner.

Figure 22

matter how hard they tried, Hoyle and his collaborators did not manage to account for the abundances of the lightest elements by forming them inside stars. Deuterium, lithium, beryllium, and boron were just too fragile—the heat in stellar interiors was sufficient for these elements to be destroyed by nuclear reactions, rather than created. Helium, the second most abundant element in the cosmos, proved to be problematic too. This may sound surprising, since stars are clearly forming helium. After all, isn't the fusion of four hydrogens into helium the main source of power for most Sun-like stars? The difficulty turned out to be not at all with synthesizing helium in general, but with synthesizing enough of it. Detailed calculations have shown that nucleosynthesis in stars would predict for helium a cosmic abundance of only about 1 percent to 4 percent, while the observed value is about 24 percent. This left the big bang as the lone source for the lightest elements, just as Gamow and Alpher had suggested.

You may have noticed that the story of the genesis of the elements—the "history of matter," as Hoyle called it—contains in it some sort of "cosmic compromise." Gamow wanted all the elements to have been created within a few minutes following the big bang ("in less time than it takes to cook a dish of duck and roast potatoes"). Hoyle wanted all the elements to be forged inside stars during the long process of stellar evolution. Nature chose a give-and-take: Light elements such as deuterium, helium, and lithium were indeed synthesized in the big bang, but all the heavier elements, and in particular those essential for life, were cooked in stellar interiors.

Hoyle got a chance to present his version of the history of matter even at the Vatican. Just a few months before the B^{2}FH paper appeared in print, the Pontifical Academy of Sciences and the Vatican Observatory organized a scientific meeting on "Stellar Populations" at the Vatican. The two dozen invitees included some of the most distinguished scientists in astronomy and astrophysics at the time. Both Fowler and Hoyle presented their results on the synthesis of the elements, and Hoyle was also asked to give a summary of the entire meeting from a physical point of view. The Dutch astronomer Jan Oort summarized from an astronomical perspective. At the opening of the meeting, on May 20, 1957, the participants met Pope Pius XII. Figure 23 shows Hoyle shaking the Pope's hand. Willy Fowler (with his back to us) stands to Hoyle's right, and Walter Baade (facing us) is to the Pope's right.

The rest, as they say, is history. The experimental and theoretical program at Kellogg Lab became, under Willy Fowler's dynamic leadership, the hub for nuclear astrophysics. Fowler went on to win the Nobel Prize in physics in 1983 (together with astrophysicist Subramanyan Chandrasekhar). Many people, including Fowler himself, felt that Hoyle should have also shared the prize. In 2008 Geoffrey Burbidge went so far as to say, "The theory of stellar nucleosynthesis is attributable to Fred Hoyle alone, as shown by his papers in 1946 and 1954 and the collaborative work of B²FH. In writing up B²FH, all of us incorporated the earlier work of Hoyle."

Why then wasn't the Nobel Prize awarded to Hoyle? Opinions vary. Geoff Burbidge concluded, based on private correspondence,

Figure 23

that a major reason for the exclusion was a perception (which he insisted was unjustified) that Fowler was the leader of B²FH. Hoyle himself apparently thought that he was denied the prize because of his criticism of the Nobel committee when it decided to award the Nobel Prize for the discovery of pulsars to Antony Hewish instead of to his graduate student Jocelyn Bell, who actually made the discovery. Others thought that Hoyle's insistence on unorthodox views concerning the big bang, which we shall discuss in detail in the next chapter, might have played a role in his not getting the prize.

What were those dissenting views? What was the background for Hoyle's opposition to the big bang?

During the years of World War II, Hoyle found himself working at the Admiralty Signals Establishment in Witley, Surrey. There he befriended two of his younger colleagues, Hermann Bondi and Thomas "Tommy" Gold, both of them Austrian-born Jews who'd escaped to England following the rise of Nazism. Ironically, prior to their work for the navy at Witley, the British government had interned the two men as enemy aliens because of their Austrian roots.

This is how Gold described his initial impression of Hoyle: "He seemed so strange; he seemed never to listen when people were talking to him, and his broad North Country accent seemed quite out of place." Very quickly, however, his opinion changed:

> I also discovered that I had misinterpreted Hoyle's attitude of apparently not listening. In fact, he listened very carefully and had an extremely good memory, as I would find out later when he frequently had remembered what I had said much better than myself. I think he put on this air not to say, "I am not listening," but instead "don't try to influence me, I am going to make up my own mind."

In their spare moments at the naval radar research facility, the trio of Hoyle, Bondi, and Gold started to discuss astrophysics, and these collaborative exchanges continued after the war. In 1945 all three returned to Cambridge, and until 1949 they spent a few hours together every day at Bondi's place. It was during that period that they started thinking about *cosmology*—the study of the entire observable universe, all treated as one entity. The Royal Astronomical Society asked Bondi to write what was then called a note, which was really a review article that would bring together an extensive body of knowledge. Hoyle suggested cosmology as the topic, since in his view "the subject had been in abeyance for a long time." To bring himself up to speed on the subject, Bondi immersed himself in the existing literature, including a sweeping 1933 article entitled "Relativistic Cosmology" by physicist Howard Percy Robertson. Hoyle, who had previously read the article, also decided to go through it again in more detail. They both realized that the almost encyclopedic essay rather dispassionately covered various possibilities for cosmic evolution without offering an opinion. In his typical nonconformist fashion, Hoyle immediately started thinking, "Has he [Robertson] really thrown his net wide enough? Are there any other possibilities?" At the same time, Gold was thrusting himself into more philosophical ideas about the universe. These were the seeds for the theory of *steady state cosmology,* which was put forward in 1948. As we shall soon discover, the theory had been a serious contender to the big bang for more than fifteen years before it became the focus of often-acrimonious controversy.

CHAPTER 9

THE SAME THROUGHOUT ETERNITY?

Bold ideas, unjustified anticipations, and speculative thought are our only means of interpreting nature . . . Those among us who are unwilling to expose their ideas to the hazard of refutation do not take part in the scientific game.

—KARL POPPER

Fred Hoyle's most enduring works were in the areas of nuclear astrophysics and stellar evolution. Yet most of those who remember him from his popular books and prominent radio programs know him as a cosmologist and co-originator of the idea of a steady state universe. What does being a cosmologist really mean?

The question "How close is the nearest planet to Earth?" is not a question in modern cosmology. Even a question on a larger scale, such as "What is the distance from the Milky Way to its nearest galaxy neighbor?" is not considered to be a question in cosmology. Cosmology deals with the average properties of our observable universe—the ones you obtain once you smooth out, over the range that our most powerful telescopes can reach. Even though galaxies tend to reside in small groups or in rich clusters, both held together by the force of gravity, once we sample large enough volumes, the universe appears to be very homogeneous and isotropic. In other words, there is no privileged position in the universe, and things look the same in all directions. Statistically speaking, any cosmic cube with a side of five hundred million light-years or larger would look roughly the same in terms of its contents, irrespective of its location in the universe. (One light-year is the distance light travels in one year, or about six trillion miles.) This broad-brush homogeneity becomes increasingly more accurate the larger the scale, up to the "horizon" of our telescopes. Cosmology deals with precisely those questions that would yield the same answer independent of the galaxy we happen to be in or the direction in which we happen to point our telescope.

Einstein had introduced the assumption of the large-scale homogeneity and isotropy of space in 1917, but this simplifying conjecture was elevated to the status of a fundamental principle in a paper published in 1933 by the English astrophysicist Edward Arthur Milne. Milne called his principle the "extended principle of relativity," requiring that "not only the laws of nature but also the events occurring in nature, the world itself, must appear the same to all observers, wherever they be." Today the stipulation of homogeneity and isotropy is known as the cosmological principle (a name coined by the German astronomer Erwin Finlay-Freundlich), and the most powerful direct evidence for its validity comes from observations of the "afterglow of creation": the cosmic microwave background radiation. This radiation is a relic of the primeval hot, dense, and opaque fireball. It comes from all directions and is isotropic to better than one part in ten thousand. (In the words of astronomer Bob Kirshner: "much smoother than a baby's bottom.)" Large-scale galaxy surveys also indicate a high degree of homogeneity. In all surveys that encompass a large enough slice of the cosmos to constitute a "fair sample," even the most conspicuous structural features are dwarfed and smoothed out.

Since the cosmological principle has proven to be so effective when applied to different positions in space, it was only natural to wonder whether it could be extended to apply to *time* as well. That is, could one argue that the universe is unchanging in its large-scale appearance as well as in its physical laws? This was the big question raised by Hoyle, Bondi, and Gold in 1948. Amusingly enough, the illustrious trio may have been inspired to ask this question by a British horror film called *Dead of Night*. (Figure 24 shows the original poster of the film.) Here is how Hoyle himself described the sequence of events:

> In a sense, the steady-state theory may be said to have begun on the night that Bondi, Gold, and I patronized one of the cinemas in Cambridge . . . It [the film *Dead of Night*] was a sequence of four ghost stories, seemingly disconnected as told by the several characters in the film, but with the interesting property that the end of the fourth story connected unexpectedly with the beginning of the first, thereby setting-up the potential for a never-ending cycle.

When the three colleagues returned to Trinity College, Gold asked suddenly, "What if the universe is like that?" meaning that the

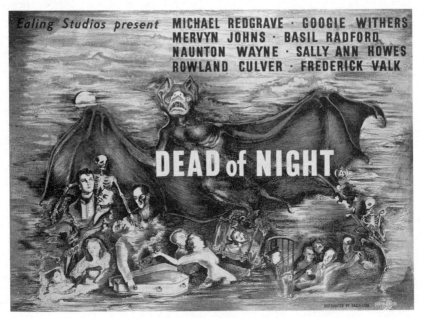

Figure 24

universe could be eternally circling on itself without a beginning or an end. The idea was certainly intriguing, except that at first blush it appeared to be at odds with the discovery by the Belgian priest and cosmologist Georges Lemaître and astronomer Edwin Hubble that the universe was expanding. The cosmic expansion seemed to be pointing rather to a linear evolution, starting from a dense and hot beginning (the big bang) and indicating a clear direction for the arrow of time. Hoyle, Bondi, and Gold were fully aware of these findings, since Hubble's discovery and its potential implications had already featured frequently in the discussions of the trio. In an interview in 1978, Gold reminisced about those intense analyses:

> What happened was that there was a period when Hoyle and I would sit around in Bondi's rooms in college a substantial amount of the time and discuss, as Hoyle always insisted, what does the Hubble thing really mean? . . . all those galaxies, all this flying apart, would the space be terribly empty afterwards? Has it been very dense in the past?

All of those contemplations had led to an unexpected outcome: Hoyle, Bondi, and Gold started to think seriously about the problem of whether the observed cosmic expansion could somehow be accommodated in the context of a theory of an unchanging universe.

But before delving into that fascinating topic, let's go back to the 1920s for a moment. The discovery of the expanding universe is not only the greatest astronomical discovery of the twentieth century, it plays such a crucial role both in Hoyle's blunder and Einstein's that it would be instructive to take a short detour to review the history of this breakthrough. This story is especially pertinent, since a new, very intriguing twist in the chronicle of events created a huge buzz in the astronomical and history of science communities in 2011.

Cosmic Expansion: Lost
(in Translation) and Found

When cosmologists say that our universe is expanding, they base this statement primarily on evidence that comes from the apparent motion of galaxies. A highly simplified, oft-used example can help visualize the concept.

Imagine a two-dimensional world that exists only on the surface of a rubber sphere (figure 25). That is, galaxies in this world are simply small, round chads (like those created with a hole punch) glued on the surface. Neither the inside of the sphere nor the space outside it exists for the inhabitants of this world; their entire universe is just the surface. Note that this world has no center; no chad on the sur-

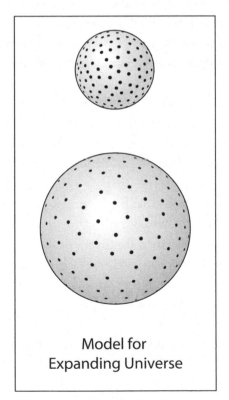

Model for
Expanding Universe

Figure 25

face is different from any other chad. (Remember that the center of the sphere itself is not a part of this world.) This universe also has no boundary or edge. If a point were to move in a certain direction on the spherical surface, it would never reach an edge.

Now, what would happen if this sphere were being inflated? Irrespective of which chad on the surface you happen to belong to, you will see all the other chads receding and rushing away from you. Moreover, chads that are more distant will be receding faster: A chad that is twice as far as another will be moving twice as fast (since it will cover twice the distance during the same period of time). In other words, the speed of recession will be proportional to the distance. Einstein's theory of general relativity articulates that the fabric of space-time (the combination of space and time into a single continuum) in our universe behaves in such a way that we can turn around this simplified example. That is, the discovery that all the distant galaxies are receding from us, combined with the fact that the speed of recession is proportional to the distance, imply that space in our universe is stretching. (We shall return to this topic in chapter 10.) Note that the expansion of the universe cannot be compared with an exploding hand grenade. In the latter case, the explosion occurs within a preexisting space, and it has a definite center (and an edge). In the universe, the receding motion arises because the fabric of space itself is stretching. No galaxy is any different from any other galaxy; from every location, you will see all the other galaxies rushing away in all directions.

The key figure with whom the discovery of cosmic expansion is usually associated is astronomer Edwin Hubble, after whom the Hubble Space Telescope was named. Hubble is commonly credited with having measured (in collaboration with his assistant, Milton Humason) the distances and recession velocities to a few dozen galaxies, and having established, in a paper published in 1929, the law that bears his name, stating that galaxies recede from us at speeds that are proportional to their distance. From that "Hubble's law," Hubble and Humason derived an overall current expansion rate suggesting that with every 3.26 million light-years of distance, the recession speed of the galaxies increases by about 500 kilometers per second, or about 311 miles per second.

Given the relatively small-distance range of Hubble's original observations, it would have been a real leap of faith to infer from them a universal expansion were it not for some supporting theoretical ideas, a few of which had even preceded the observations. In fact, as early as 1922, the Russian mathematician Aleksandr Friedmann showed that general relativity allowed for an expanding, matter-filled, unbounded universe. While few took notice of Friedmann's results (other than Einstein himself, who eventually acknowledged their mathematical correctness but dismissed them, since he thought that "a physical significance can hardly be ascribed to them"), the notion of a dynamical universe was starting to gain influence during the 1920s. Consequently, the interpretation of Hubble's observations in terms of an expanding universe became popular fairly fast.

Physicists sometimes tend to ignore the history of their subject. After all, who cares who discovered what as long as the discoveries are made widely known. Only totalitarian regimes have been obsessed with insisting that all good ideas are homegrown. In an old joke about the Soviet Union, an important visitor is brought to the science museum in Moscow. In the first room, he sees a giant picture of a Russian man he had never heard of. When he asks who that person is, he is told, "This is so-and-so, the inventor of the radio." In the second room: another giant portrait of a complete stranger. "The inventor of the telephone," his host informs him. And so it continues for about a dozen rooms. In the final room, there is a picture that dwarfs by comparison all of the other pictures. "Who is this?" the visitor asks in astonishment. The host smiles and answers, "This is the man who invented all of those other men in the previous rooms."

In a few cases, however, discoveries are of such magnitude that understanding the path that had led to these insights—including the correct attribution—can be of great value. There is very little doubt that the discovery of the expansion of the universe falls into this category, even if for no other reason than the fact that the expansion suggests that our universe had a beginning.

During 2011, a passionate debate flared up about who actually deserves the credit for discovering the cosmic expansion. In particular, a few articles even raised the suspicion that some improper censorship practices may have been applied in the 1920s to ensure Edwin Hubble's priority on the discovery.

Here are, very briefly, the background facts that are most relevant for this debate.

By February 1922, astronomer Vesto Slipher had measured the radial velocities (velocities along the line of sight from us) for forty-one galaxies. In a book published in 1923, Arthur Eddington listed those velocities and remarked, "The great preponderance of positive [receding] velocities is very striking; but the lack of observations of southern nebulae is unfortunate, and forbids a final conclusion." (Galaxies were initially called nebulae [from Latin for "mist," or "cloud"] because of their fuzzy appearance.) In 1927 Georges Lemaître published (in French) a remarkable paper whose title read (in its English translation): "A Homogeneous Universe of Constant Mass and Increasing Radius Accounting for the Radial Velocity of Extra-Galactic Nebulae." Unfortunately, it was published in the little-read *Annals of the Brussels Scientific Society*. In it, Lemaître first discovered dynamic (expanding) solutions to Einstein's general relativity equations, from which he derived the theoretical basis for what is now known as Hubble's law: the fact that the velocity of recession is directly proportional to the distance. But Lemaître went beyond mere theoretical calculations. He actually used the velocities of the galaxies as measured by Slipher—and approximate distances as determined from brightness measurements by Hubble in 1926—to discover the existence of a tentative "Hubble's law" and to determine the rate of expansion of the universe. For the numerical value of that rate, today called the Hubble constant, Lemaître obtained 625 (in the common units of kilometers per second for every 3.26 million light-years of distance). Two years later, Edwin Hubble obtained a value of about 500 for this same quantity. (Both values are known today to have been wrong by almost an order of magnitude.) In fact, Hubble used essentially the same recession velocities—the ones determined by Slipher—without ever mentioning in his paper that these were the latter's work. Hubble did use superior distances, which were based in part on better stellar distance indicators. Lemaître was fully aware of the fact that the distances he had used were only approximate. He concluded that the accuracy of the distance estimates available at the time seemed insufficient to assess the validity of the linear relation he had discovered.

Based solely on what I have described so far, I think most people would agree that it seems only fair to attribute the discovery of the expanding universe and of the tentative existence of Hubble's law to Lemaître, and the detailed confirmation of that law to Hubble and Humason. The subsequent, truly meticulous observations of Hubble and Humason extended Slipher's velocity measurements to greater and much more accurate distances. Here, however, is where the plot thickens.

The English translation of Lemaître's 1927 paper was published in the *Monthly Notices of the Royal Astronomical Society* in England in March 1931. However, a few paragraphs from the original French version were deleted—in particular, the paragraph that described Hubble's law and in which Lemaître used the forty-two galaxies for which he had (approximate) distances and velocities to derive a value for the Hubble constant of 625. Also missing were one paragraph in which Lemaître discussed the possible errors in the distance estimates, and two footnotes, in one of which he remarked on the interpretation of the proportionality between the velocity and distance as resulting from a relativistic expansion. In the same footnote, Lemaître also calculated two possible values for the Hubble constant: 575 and 670, depending on how the data were grouped.

Who translated the article? And why were these paragraphs deleted from the English version? Several history-of-science amateur sleuths suggested in 2011 that someone had deliberately censored those parts of Lemaître's paper that dealt with Hubble's law and the determination of the Hubble constant. Canadian astronomer Sidney van den Bergh speculated that whoever did the "selective editing" did so to prevent Lemaître's paper from undermining

Edwin Hubble's priority claim. "Picking out part of the middle of an equation must have been done on purpose," he noted. South African mathematician David Block went even somewhat further. He suggested that Edwin Hubble himself might have had a hand in this cosmic "censorship" to ensure that credit for the discovery of the expanding universe would go to himself and the Mount Wilson Observatory, where he made the observations.

As someone who has worked for more than two decades with Hubble's namesake—the Hubble Space Telescope—I became sufficiently intrigued by this whodunit to attempt to appraise the facts more carefully. I started by examining the circumstances surrounding the translation of Lemaître's article.

First, I obtained a copy of the original letter sent by the editor of the *Monthly Notices* at the time, astronomer William Marshall Smart, to Georges Lemaître. In that letter (figure 26), Smart asked Lemaître whether he would allow his 1927 paper to be reprinted in the *Monthly Notices,* since the Royal Astronomical Council felt that the paper was not as well known as its importance deserved. The most important paragraph in the letter reads:

> Briefly—if the Soc. Scientifique de Bruxelles [in the annals of which the original paper was published] is also willing to give its permission—we should prefer the paper translated into English. Also, if you have any further additions etc. on the subject, we would glad[ly] print these too. I suppose that if there were additions a note could be inserted to the effect that §§1–n are substantially from the Brussels paper + the remainder is new (or something more elegant). Personally and also on behalf of the Society I hope that you will be able to do this.

My immediate reaction was that the text of Smart's letter was entirely innocent, and it certainly did not suggest any intent of extra editing or censorship. But while I was fairly convinced of the correctness of this nonconspiratorial interpretation of Smart's letter, the two main mys-

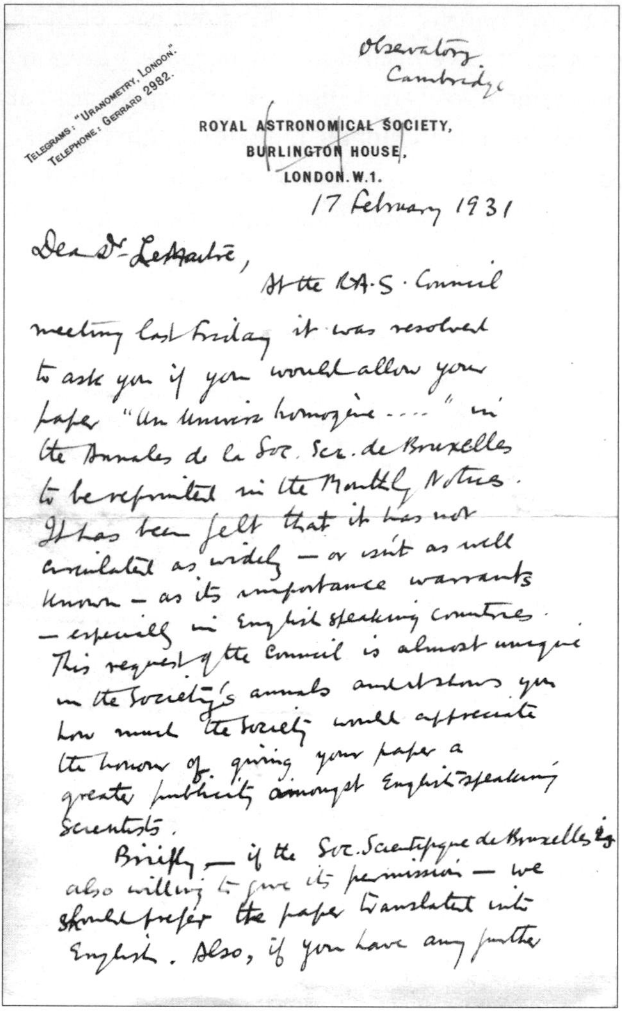

Figure 26a

teries—who translated the paper and who deleted the paragraphs— remained unresolved. In an attempt to answer these questions definitively, I decided to explore the matter further by scrutinizing all of the council's minutes and the entire surviving correspondence from 1931 at the Royal Astronomical Society Library in London. After going through many hundreds of irrelevant documents and almost giving up, I discovered two "smoking guns." First, in the minutes of

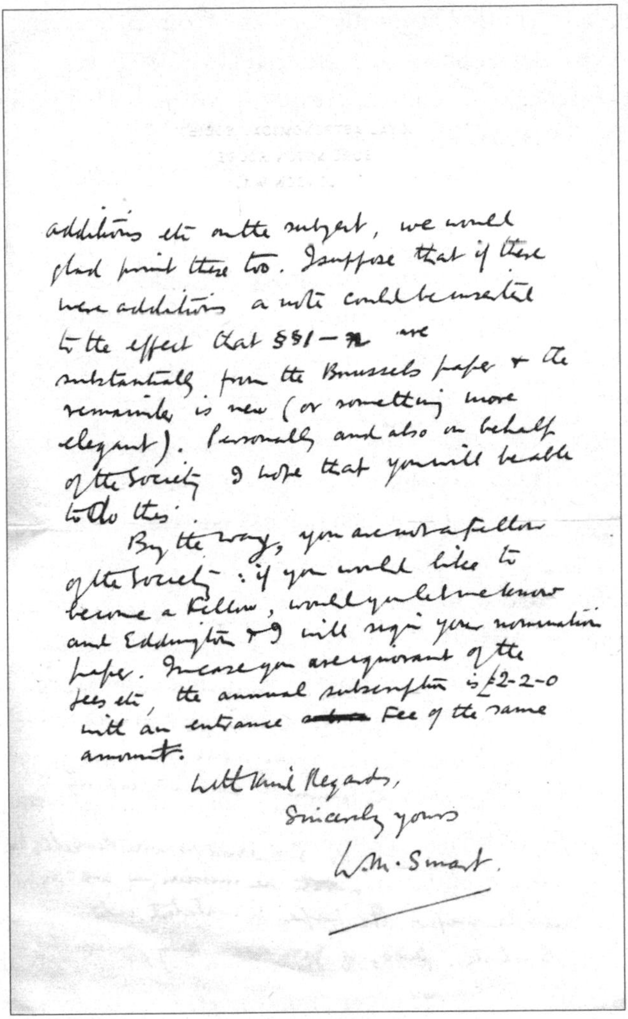

Figure 26b

the council from February 13, 1931, it is reported: "On the motion of Dr. Jackson it was resolved that the Abbé Lemaître be asked if he would allow his paper 'Un Univers homogène de masse constante et de rayon croissant,' or an English translation thereof, to be published in the Monthly Notices." This, of course, was precisely the decision mentioned in Smart's letter to Lemaître. (As an amusing aside, the same minutes also report, "A motion by Sir Arthur Eddington that smoking be permitted at meetings of the Council was discussed. It was resolved that smoking be permitted after 3:30 p.m.") The second piece of evidence was Lemaître's response to Smart's letter (figure 27), dated March 9, 1931. The letter reads:

Dear Dr. Smart

I highly appreciate the honour for me and for our society to have my 1927 paper reprinted by the Royal Astronomical Society. *I send you a translation of the paper. I did not find advisable to reprint the provisional discussion of radial velocities which is clearly of no actual interest* [Lemaître almost certainly was translating the French word *actuel,* which means "current"], *and also the geometrical note, which could be replaced by a small bibliography of ancient [old] and new papers on the subject* [emphasis added]. I join a french text with indication of the passages omitted in the translation. I made this translation as exact as I can, but I would be very glad if some of yours would be kind enough to read it and correct my english which I am afraid is rather rough. No formula is changed, and even the final suggestion which is not confirmed by recent work of mine has not be modified. I did not write again the table which may be printed from the french text.

As regards to addition on the subject, I just obtained the equations of the expanding universe by a new method which makes clear the influence of the condensations and the possible causes of the expansion. I would be very glad to have them presented to your society as a separate paper.

I would like very much to become a fellow of your society and would appreciate to be presented by Prof. Eddington and you.

If Prof. Eddington has yet a reprint of his May paper in M.N. I would be very glad to receive it.

Will you be kind enough to present my best regards to professor Eddington.

This clearly puts to bed all the speculations about who translated the paper and who deleted the paragraphs: Lemaître himself did both!

Lemaître's letter also provides a fascinating insight into the scientific psychology of (at least some of) the scientists of the 1920s. Lemaître was not at all obsessed with establishing priority for his original discovery. Given that Hubble's results had already been published in 1929, he saw no point in repeating his more tentative earlier findings again in 1931. Rather, he preferred to move forward

Louvain, le 9 mars 1931

Dear Dr. Smart

I highly appreciate the honour for me and for our society to have my 1927 paper reprinted by the Royal Astronomical Society. I send you a translation of the paper. I did not find advisable to reprint the provisional discussion o.f radial velocities which is clearly of no actual interest, and also the geometrical note, which could be replaced by a small bibliography af ancient and new papers on the subject. I join a french text with indication of the passages omitted in the translation. I made this translation as exact as I can, but I would be very glad if some of yours would be kind enough to read it and correct my english which I am afraid is rather rough. No formula is changed, and even the final suggestion which is not confirmed by recent work of mine has not be modified. I did not write again the table which may be printed from the french text.

As regards to addition on the subject, I just obtained the equations of the expanding universe by a new method which makes clear the influence of the condensations and the possible causes of the expansion. I would be very glad to have them presented to your society as a separate paper.

I would like very much to become a fellow of your society and would appreciate to be presented by Prof. Eddington and you.

If Prof. Eddington has yet a reprint of his May paper in M.N. I would be very glad to receive it.

Will you kind enough to present my best regards to professor Eddington

    and beleive

        yours sincerely

        S. Lemaître

        40 rue de Namur
        Louvain

Figure 27

and to publish his new paper, "The Expanding Universe," which he did. Lemaître's request to join the Royal Astronomical Society was also granted eventually. Lemaître was officially elected as an associate on May 12, 1939.

The Steady State Universe

Returning now to Gold's provocative question "What if the universe is like that?"—referring to the circular plot of the film *Dead of the Night*—the possibility was not considered palatable by his two colleagues; at least not initially. Hoyle immediately brushed off Gold, scoffing, "Ach, we shall disprove this before dinner." This "prediction," however, turned out to be wrong. In Bondi's words, "Dinner was a little late that night, and before very long we all said that this was a perfectly possible solution." In other words, a never-changing universe, with no beginning and no end, started to look more and more attractive. From that point on, however, Hoyle took a somewhat different approach to the problem from that of his scientific peers.

The outlook of Bondi and Gold was based on an appealing philosophical concept. If the universe is indeed evolving and changing, they argued, then there is no clear reason why we should trust that the laws of nature have permanent validity. After all, those laws were established based on experiments performed here and now. In addition, Bondi and Gold perceived that the cosmological principle, as originally stated, presented yet another difficulty. It assumed that observers located in different galaxies anywhere in the universe would all discern the same large-scale picture of the cosmos. But if the universe is continuously evolving with time, this required that the different observers would compare their notes at the same time, which implied that one needed to define what precisely is meant by "at the same time." To circumvent all of these obstacles, Bondi and Gold proposed their Perfect Cosmological Principle, which added to the original principle the requirement that there is no preferred time in the cosmos—the universe looks the same from every point *at all times.*

Even though Hoyle decided to take a different route, he did find this intuitive principle of Bondi and Gold compelling, especially since it also solved another problem inferred from the observations of the expanding universe. Hubble's determination of the rate of expansion (which was later found to be wrong) implied a nightmarish scenario in which the universe was only 1.2 billion years old—far less than the estimated age of the Earth! So in spite of Hubble's enormous prestige ("more than life sized in the 30s and 40s" according to Bondi), Hoyle, Bondi, and Gold felt that another solution had to be found. Unlike Bondi and Gold, however, Hoyle embarked on a more mathematical, rather than philosophical, approach. In particular, he developed his theory in the framework of Einstein's general relativity. He started from the observational fact that the universe is expanding. This immediately raised a question: If galaxies are continuously rushing away from each other, does that mean that space is becoming more and more empty? Hoyle answered with a categorical no. Instead, he proposed, matter is continually being created throughout space so that new galaxies and clusters of galaxies are constantly being formed at a rate that compensates precisely for the dilution caused by the cosmic expansion. In this way, Hoyle reasoned, the universe is preserved in a steady state. He once commented wittily, "Things are the way they are because they were the way they were." The difference between the steady state universe and the evolving (big bang) universe is shown schematically in figure 28, where I have again used the analogy of the inflating sphere. In both cases, we start (at the top) with a sample of the universe, in which the galaxies are represented by small round chads. In the evolutionary scenario (on the left), after some time has passed, the galaxies have receded from one another (bottom left), reducing the overall density of matter. In the steady state scenario, new galaxies have been created, so that the average density remained the same (bottom right).

The idea of matter being continuously created out of nothing may appear crazy at first. However, as Hoyle was quick to point out, no one knew where matter had appeared from in the big bang cosmology, either. The only difference, he explained, was that in the

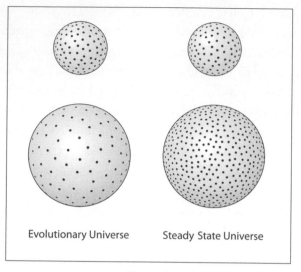

Evolutionary Universe   Steady State Universe

Figure 28

big bang scenario all the matter was created in one explosive beginning, while in the steady state model matter has been created at a constant rate throughout an infinite time and is still being created at the same rate today. Hoyle contended that the concept of continuous creation of matter (when put in the context of a specific theory) was much more attractive than creation of the universe in the remote past, since the latter implied that observable effects had arisen from "causes unknown to science." To achieve a steady state, Hoyle added to Einstein's general relativity equations a "creation field" term, the effect of which was to create matter spontaneously. What sort of matter? Hoyle did not know for sure, but he conjectured, "Neutron creation appears to be the most likely possibility. Subsequent disintegrations might be expected to supply the hydrogen required by astrophysics. Moreover, the electrical neutrality of the universe would then be guaranteed." The rate at which new atoms were supposed to materialize out of empty space was too small to be directly observable. Hoyle described it once as "about one atom every century in a volume equal to the Empire State Building."

The key virtue of the steady state scenario was that, as expected from all good scientific theories, it was falsifiable. Here is how philosopher of science Karl Popper expressed his views on what constitutes a theoretical system of natural science:

> I shall not require of a scientific system that it shall be capable of being singled out, once and for all, in a positive sense; but I shall require that its logical form shall be such that it can be singled out, by means of empirical tests, in a negative sense: *it must be possible for an empirical scientific system to be refuted by experience.*

The steady state model predicted that galaxies that are billions of light-years away should look, statistically speaking, just like nearby galaxies, even though we see the former as they were billions of years ago because of the time it takes their light to reach us. Bondi used to challenge the supporters of the evolving universe (big bang) model by saying, "If the universe has ever been in a very different state from what it is now, show me some fossil remains of what it was like a long time ago." In other words, if, for instance, extremely remote galaxies were found to look (on the average) very different from galaxies in the neighborhood of the Milky Way, our universe could not be in a steady state.

Evolution

When Hoyle, and, separately, Bondi and Gold, published their steady state papers, they presented the astrophysics community with a choice between two very different world views. On one hand, there was the big bang model, in which the universe was assumed to have had a beginning in the form of a dense and hot state (which Lemaître called the "primeval atom"). In addition to Lemaître, George Gamow was perhaps the strongest advocate for this scenario. As we saw in the last chapter, Gamow even (mistakenly) thought that all the chemical elements had been forged in this cosmic initial explosion.

In contrast to the big bang stood the steady state model, with its infinite past and unchanging cosmic scenery, despite the overall expansion. However, the telescopes of the late 1940s were not powerful enough to detect whether an evolutionary trend of the type implied by the big bang model existed or not. When Hoyle met Edwin Hubble for the first time, in August of 1948, he was delighted to hear from the latter that what was supposed to become the world's largest telescope—the two-hundred-inch telescope on Mount Palomar in California—was undergoing its final testing. Hubble hoped to start observing remote galaxies soon thereafter. Disappointingly, however, even the large mirror of the Mount Palomar telescope could not collect enough light from very distant, ordinary galaxies to distinguish unambiguously between the two rival theories.

In October 1948, Hoyle, Bondi, and Gold attended a small meeting of the Royal Astronomical Society in Edinburgh. All three of them were invited to present their ideas about the steady state universe. Hoyle used the opportunity to advance for the first time a possible connection between an unchanging, self-sustaining cosmos and life:

> Modern astrophysics appears to be inexorably forcing us away from a universe of finite space and time, in which the future holds nothing but a general running down or heat death, towards a universe in which both space and time are infinite. The possibilities of physical evolution, and perhaps even of life, may well be without limit. These are the issues that stand to-day before the astronomer. Within a generation we hope that they can be settled with reasonable certainty.

Paradoxically, even though later in life Hoyle criticized natural selection (claiming a role for panspermia, or life as a cosmic phenomenon), the origin of this line of thinking could be traced back to Darwin. Recall that Darwin was concerned about Kelvin's estimate of the age of the Earth because he feared that with the restricted age there wasn't sufficient time for evolution to operate. Hoyle here alludes to an advantage of the steady state theory: A universe that has always existed and will exist forever affords an infinite amount of time for life to emerge and to evolve. We shall return to this question later, when we'll discuss the possible reasons for Hoyle's stubborn clinging to the steady state idea.

Following the presentations by Gold, Bondi, and Hoyle, the president of the Royal Astronomical Society, astronomer William Greaves, opened the subsequent discussion with a somewhat sarcastic remark: "Cosmology is one department of astronomy—sometimes I suspect its adherents of thinking it is the only part—but we all agree that it is a most important part." As it so happened, one of the most distinguished physicists of the twentieth century, Max Born, was in attendance. When asked for his reaction to the steady state model, Born said:

> I am overawed by the whole character of the cosmologists! After the initial discoveries of atomic physics, physicists continue to find new particles at frequent intervals: so in cosmology we shall continue to discover new theories of world structure and evolution . . . I am filled with gratitude at hearing these papers, but I am skeptical.

The first signs of trouble for the steady state model came not from optical telescopes but from radio astronomy. The universe is essentially transparent to radio waves, and, consequently, the antennae of radio telescopes could pick up signals even from distant (but "active" in the radio spectral range) galaxies that could barely be detected optically. In the 1950s, British and Australian scientists put to good use the expertise gained during World War II to develop a strong radio astronomy program. One of the pioneers in this endeavor was a physicist from the Cavendish Laboratory at Cambridge: Martin Ryle.

Unlike Hoyle, Ryle came from a privileged background—his father was physician to King George VI—and he had received the best of what private education could offer. After some pioneering radio observations of the Sun in the late 1940s, Ryle and his group embarked on an ambitious program to detect radio sources beyond the solar system. Following some impressive improvements to the observational techniques that allowed them to discard background radiation from the Milky Way, Ryle and his colleagues discovered several dozen "radio stars" distributed more or less isotropically across the sky. Unfortunately, since most of the sources did not have visible counterparts, there was no way to determine their distances precisely. Ryle was of the opinion that these were peculiar stars within our own galaxy, and he was prepared to forcefully defend this view at a small gathering of radio astronomy enthusiasts.

This so-called Massey Conference (named after atomic physicist Harrie Massey, who hosted it) took place at University College London in March 1951. Both Hoyle and Gold were present, and they did not hide their skepticism. At one point, Gold stood up and challenged Ryle's conclusions. He contended that since the discrete radio sources were uniformly distributed in all directions, rather than being concentrated toward the plane of the Milky Way, they must be outside our own galaxy, at much larger distances. The only alternative, he argued, was that the sources were in fact so close that they were all contained within the relatively small thickness of the Galactic disk (distances shorter than one hundred light-years). Ryle's hypothesis, that the sources were scattered all across the Milky Way, was untenable in Gold's view. Hoyle fully supported Gold's position, provoking a sarcastic comment from Ryle: "I think the theoreticians have misunderstood the experimental data." Hoyle responded by pointing out that of the half a dozen sources or so that had actually been optically identified, five corresponded to external galaxies. Years later, he commented that Ryle used the word "theoreticians" in a way that implied some "inferior and detestable species."

This was but one of the many major clashes between the steady state theorists and Ryle, and it left emotional scars on both Hoyle and Ryle. In this particular case, Gold and Hoyle prevailed.

About a year after the Massey meeting, astronomer Walter Baade determined that the distance to a radio source in the constellation

Cygnus was hundreds of millions of light-years, confirming Hoyle's suspicion. Ironically, however, it was precisely the great distance of the radio sources that later became the cornerstone of Ryle's argument in favor of an evolving universe and which led to the downfall of the steady state theory. (The steady state theory never created much resonance in the United States, but in 1952, following a lecture by the Astronomer Royal, Sir Harold Spencer Jones, it did manage to generate a few headlines. Two of these, one in the *New York Times* and the other in the *Christian Science Monitor*, are shown in figure 29.)

Ryle had to suffer one more temporary embarrassment in his campaign against steady state cosmology, even though that particular sequence of events started with what had appeared to be a victory. The big bang and steady state models made distinctly different predictions about the distant universe. When we observe galaxies that are billions of light-years away, we get a picture of those galaxies as they *were* billions of years ago. In a continuously evolving universe (the big bang model), this means that we observe that particular part of the universe when it was younger and therefore different. In the steady state model, on the other hand, the universe has always existed in the same state. Consequently, the remote parts of the universe are expected to have precisely the same appearance as the local cosmic environment. Ryle seized on the opportunity afforded by this testable prediction and started to collect a large sample of

Figure 29

radio sources, and to count how many of them there were at different intensity intervals. Since he had no way of knowing the actual distances to most sources (they were beyond the detection range of optical telescopes), Ryle made the simplest assumption: namely, that the observed weaker radio sources were, on average, more distant than the sources of the strong signals. He found that there were dramatically more weak sources than strong ones. In other words, it seemed that the density of sources at distances of billions of light-years (and therefore representing the universe billions of years ago) was much higher than the current density nearby. This was clearly at odds with a model of a never-changing universe, but it could be made consistent with a cosmos evolving from a big bang, if one assumed (correctly, as we now recognize) that galaxies were more prone to emit intense radio signals in their youth than at present, in their older age.

Ryle presented his results on May 6, 1955, when he was giving the prestigious Halley Lecture (named after the famous seventeenth-century astronomer Edmond Halley). Without ever mentioning Hoyle by name, referring only to "Bondi and others" as the originators of the steady state model, Ryle's verdict was unambiguous: "If we accept the conclusion that most of the radio stars are external to the Galaxy, and this conclusion seems hard to avoid, then there seems to be no way in which the observations can be explained in terms of a steady state theory."

Ryle continued his attack a week later, when at the May 13 meeting of the RAS, he and his student John Shakeshaft were delighted to close by saying, "We must conclude that the remote regions of the Universe differ from those in our neighborhood, a result which is not compatible with steady-state cosmological theories, but which may well be accounted for in terms of evolutionary theories."

Confronted with this serious challenge, Gold and Bondi, who attended the RAS meeting, found themselves on the defensive. Gold decided to craftily remind the audience that Ryle had been wrong before. He pointed out that he was "glad to see that there is now agreement that many of these sources are likely to be extragalactic," as he himself had suggested four years earlier, when "Mr. Ryle . . . considered that such a suggestion must be based on a misunderstanding of the evidence." He then added that based on the information presented, it was "very rash to regard the great majority of weak sources as extremely distant." He cautioned that if the sources were not all the same, but, rather, there was a wide range of intensities among the intrinsic radio signals, then Ryle's counting of weak sources could represent a confusing mix of faraway sources with nearby ones. Bondi was also skeptical of the interpretation of Ryle's results. In his view, the uncertainties that still existed in the counts did not allow for conclusive inferences. To drive home this point, he reminded his audience that earlier attempts aimed at determining the geometry of the universe based on galaxy counts resulted in totally disparate conclusions.

It goes without saying that Hoyle himself did not agree with Ryle's interpretation. Rather than engaging in long arguments, however, he decided to wait for superior observational data to surface and refute Ryle's finding. To the surprise of many astronomers, such contradictory results have indeed emerged. Australian radio astronomers showed in 1957 that Ryle's earlier survey was seriously flawed: The map of radio sources that Ryle had produced was so blurred that blends of two or more radio sources were often counted as one. The consequences were clear to the Australian astronomers: "Deductions of cosmological interest derived from the analysis are without foundation."

Hoyle did not bother to rejoice. The year 1957 witnessed the publication of the celebrated B^2FH, and he was deeply engrossed in the synthesis of the elements rather than in steady state cosmology. It had not escaped him, though, that forging most of the nuclei in stellar cores (instead of in a big bang) could also be seen as supporting (at least partially) a steady state perspective. In the same year, Hoyle was also elected as Fellow of the Royal Society, an honor that put him on par with Ryle in terms of his academic status. But Ryle did not give up. He and his team continued to introduce significant upgrades both to their instrumentation and to the data reduction and

analysis. Their efforts resulted in the production of the third generation of the Cambridge catalogue of radio sources (known as the *3C
Catalogue*).

By the early 1960s, Ryle's group had at its disposal even an
entirely new radio observatory, funded by the Mullard electronics
company. The intellectual skirmishes between Ryle and Hoyle continued, culminating in one particularly unpleasant incident. Hoyle
later described this traumatic experience in his autobiographical book
Home Is Where the Wind Blows. It all started with what appeared to
be an innocent phone call from the Mullard company in early 1961.
The person at the other end of the line invited Hoyle and his wife to
attend a press conference at which Ryle was expected to present new
results that were supposed to be of great interest to Hoyle. When
they arrived at the Mullard headquarters in London, Hoyle's wife,
Barbara, was escorted to a seat in the front now, while Hoyle was
led to a chair on stage, facing the media. He had no doubt that the
announcement would be related to the counting of radio sources
according to their intensity, but he couldn't believe that he would
have been invited if the results were to contradict the steady state
theory. In his words:

> Was I being uncharitable in thinking that the new results
> Ryle would shortly be announcing were adverse to my
> position? Surely, if they were adverse, I would hardly have
> been set up so blatantly. Surely, it must mean that Ryle was
> about to announce results in consonance with the steady-
> state theory, ending with a handsome apology for his pre-
> viously misleading reports. So, I set about composing an
> equally handsome reply in my mind.

Unfortunately, what Hoyle found utterly unthinkable did happen. When Ryle appeared, rather than making a brief announcement,
as advertised, he launched into a technical, jargon-filled lecture on
the results of his larger, fourth survey. He finished by claiming con-
fidently that the results now showed unambiguously a higher den-

sity of radio sources in the past, therefore proving the steady state theory wrong. The shocked Hoyle was merely asked to comment on the results. Incredulous and humiliated, he barely mumbled a few sentences and rushed away from the event. The media frenzy that followed in the subsequent days disgusted Hoyle to the point that he avoided phone calls for a week and was absent even from the following RAS meeting on February 10. Even Ryle realized that the press conference had crossed the border of common decency. He called Hoyle to apologize, adding that when he agreed to the Mullard event, he "had no idea how bad it would be."

On the purely scientific front, however, despite these disturbing failures in etiquette, Ryle's arguments grew increasingly compelling, and by the mid-1960s, the vast majority of the astronomical community agreed that the proponents of the steady state theory had lost the battle. (Figure 30 shows, from left to right, Gold, Bondi, and Hoyle, attending a conference in the 1960s.) The discovery of extremely active galaxies, in which the accretion of mass onto central, supermassive black holes releases sufficient radiation to outshine the entire galaxy, cemented the evidence against a steady state universe. These objects, known as *quasars*, were luminous enough to be observed by optical telescopes. The observations allowed astronomers to use Hubble's law to determine the distance to these sources,

Figure 30

and to show convincingly that quasars were indeed more common in the past than at present. There was no escape from the conclusion that the universe was evolving and that it had been denser in the past. At that point, the floodgates opened, and the challenges to the steady state model kept pouring in. In particular, in 1964 scientists Arno Penzias and Robert Wilson made a discovery that to all but its diehard supporters represented the last nail in the coffin of the steady state theory.

Penzias and Wilson were working at the Bell Telephone Laboratories in New Jersey with an antenna built for communication satellites. To their annoyance, they were picking up some sort of pervasive background radio noise: microwave radiation that appeared to be the same from all directions. After failing to explain away this disturbing "hiss" as an instrumental artifact, Penzias and Wilson finally announced the detection of an intergalactic temperature excess of about 3 Kelvin (3 degrees above absolute zero). Lacking the necessary background, Penzias and Wilson did not realize initially what they had found. Robert Dicke of Princeton University, however, recognized the signal immediately. Dicke was in the process of building a radiometer to search for the relic radiation from the big bang, previously predicted by Alpher, Hermann, and Gamow. Consequently, his correct interpretation of the results of Penzias and Wilson literally transformed the big bang theory from hypothesis into experimentally tested physics. As the universe expanded, the incredibly hot, dense, and opaque fireball cooled down continuously, eventually reaching its present temperature of about 2.7 Kelvin.

Since then, observations of the cosmic microwave background have produced some of the most precise measurements in cosmology. The temperature of this radiation is now known to four significant figures to be 2.725 K, and its intensity changes with wavelength precisely as expected from a thermal source—confirming the predictions of the big bang. Even in the face of this overwhelming, contradictory evidence to the steady state theory, Hoyle was never convinced. He proposed that instead of representing a relic of the big bang, the cosmic microwave background is produced by some extra-galactic iron "whiskers," which absorb and scatter the infrared light of galaxies at microwave wavelengths. These iron whiskers were supposed to have condensed from metallic vapors—for instance, in the material ejected by supernova explosions.

In spite of Hoyle's valiant efforts, beginning in the mid-1960s most scientists stopped paying attention to the steady state theory. Hoyle's continuing attempts to demonstrate that all the confrontations between the theory and emerging observations could be explained away looked increasingly contrived and implausible. Worse yet, he seemed to have lost that "fine judgment" that he had once advocated, which was supposed to distinguish him from "merely becoming a crackpot." At an international symposium on the topic "Modern Cosmology in Retrospect," which took place in Bologna, Italy, in 1988, he gave a talk entitled "An Assessment of the Evidence Against the Steady-State Theory." In that downright anachronistic talk, Hoyle tried (unsuccessfully, I should add) to convince his audience that all the compelling pieces of evidence for the big bang—the existence of the cosmic microwave background; the implied need for a primordial synthesis of the light elements deuterium, helium, and lithium; and the counts of the radio sources—could *all* still be explained by the steady state theory. Hoyle's obstinate resistance to changing his views stood in stark contrast to the attitude adopted, for instance, by co-originator of the steady state theory Hermann Bondi. Recall that Bondi had insisted on being shown some fossil remains of what the universe was like in the past, if the universe was indeed evolving. In his own talk at the same conference in Bologna, Bondi admitted that such fossil evidence had indeed emerged, both in the form of the cosmic helium abundance, which had been shown to have most likely formed in the big bang, and in the cosmic microwave background, which beautifully matched the big bang predictions. Bondi therefore concluded graciously, "So my challenge of whether fossils could be found has had an answer long after I posed it."

Hoyle, on the other hand, continued to advocate a somewhat modified version of the steady state theory (which he called "quasi-steady state"). Even as late as the year 2000, at the age of eighty-five, he published a book entitled *A Different Approach to Cosmology: From a Static Universe Through the Big Bang Towards Reality,* in which he and his collaborators, Jayant Narlikar and Geoff Burbidge, explained the details of the quasi–steady state theory and their objections to the big bang. To express their contemptuous opinion of the scientific establishment, they presented in one of the book's pages a photograph of a flock of geese walking on a dirt road with the caption, "This is our view of the conformist approach to the standard (hot big bang) cosmology. We have resisted the temptation to name some of the leading geese." By then, however, Hoyle had been out of the conventional cosmological wisdom for so long that very few even bothered to point out the shortcomings of the modified theory. Perhaps the best thing said about the book appeared in the review by Britain's *Sunday Telegraph*, and it referred not so much to the contents of the book as to Hoyle's fiery personality: "Hoyle systematically reviews the evidence for the Big Bang theory, and gives it a good kicking . . . it's hard not to be impressed with the audacity of the demolition job . . . I can only hope that I possess one-thousandth of Hoyle's fighting spirit when I, like him, have reached my 85th year."

Dissidence and Denial

Hoyle's blunder was somewhat different from those of Darwin, Kelvin, and Pauling in two important respects. First, there was the issue of the scale of the topic, in the context of which the blunder occurred. Darwin's blunder involved only one element of his theory (albeit an extremely important one). Kelvin's blunder concerned an assumption at the basis of a particular calculation (a very meaningful one). Pauling's blunder affected one specific model (unfortunately for the most crucial molecule). Hoyle's blunder, on the other hand, concerned no less than *an entire theory* for the universe as a whole. Second and more important, Hoyle did nothing wrong in proposing the steady state model — unlike Darwin, who did not understand the implications of a faulty biological mechanism; Kelvin, who neglected unforeseen physical processes; and Pauling, who ignored basic rules of chemistry. The theory itself was bold, exceptionally clever, and it matched all the observational facts that existed at the time. Hoyle's blunder was in his apparently pigheaded, almost infuriating refusal to acknowledge the theory's demise even as it was being smothered by accumulating contradictory evidence, and in his use of asymmetrical criteria of judgment with respect to the big bang and steady state theories. What was it that caused this intransigent behavior? To answer this intriguing question, I started by asking a few of Hoyle's former students and younger colleagues for their opinions.

Cosmologist Jayant Narlikar was Hoyle's graduate student, and he continued to collaborate with him throughout Hoyle's life. The two researchers developed, among other things, a theory of gravity known as the Hoyle-Narlikar theory, which fits into their quasi–steady state model. Narlikar suggested that Hoyle's displeasure with the big bang model stemmed, at least initially, from genuine discomfort that Hoyle felt with some of the physical premises of the big bang. For instance, Narlikar recalled, Hoyle pointed out that all the other observed background radiations (optical, X-ray, infrared) were found to be associated with astrophysical objects (stars, active galaxies, and so on), and he saw no reason why the cosmic microwave background would be different and related to a singular event (the big bang). Similarly, around 1956, he thought that stars could somehow produce the energy observed in the cosmic microwave background, if one could find a way to synthesize all the helium in stars. On the more emotional side, Narlikar felt that the fact that Hoyle was not a religious person might have also contributed to his objection to a universe that appeared all at once.

Astrophysicists Peter Eggleton and John Faulkner were both Hoyle's research students in the early 1960s (Faulkner is the person farthest right in the front row in figure 22), but I was somewhat surprised to discover that their sentiments were rather different. Eggleton remembered Hoyle as a person who knew everything that was worth knowing in astrophysics at the time and also knew everybody that was anybody in the world of astrophysics. He remarked that a whimsical line that had been used to describe the Victorian scholar Benjamin Jowett could be adopted as a genuine characterization of Hoyle, namely: "What he didn't know wasn't knowledge." Concerning Hoyle's attitude toward science, Eggleton's impression was that if the scientific community believed something, Hoyle would be inclined to believe the opposite, to see how far he could go. When I pressed him on why he thought Hoyle was so reluctant to accept the big bang, Eggleton expressed the view that Hoyle's rejection of the idea that life on Earth emerged through a natural, chemical evolution was at the root of this resistance. Hoyle insisted, Eggleton said, that the origin of life required much more time than the age of the universe as inferred from the big bang theory. This is an interesting point, to which we shall return shortly.

Faulkner admitted that he himself was puzzled by Hoyle's unyielding position toward the big bang. In his opinion, Hoyle "went off the rails a bit, having developed a love for his brainchild [the steady state theory] and not wanting to give it up." He made another interesting comment that by the late 1960s, Hoyle's interest in what one might call "normative science" decayed, giving way to a more maverick path.

Martin Rees, Astronomer Royal for Britain, succeeded Hoyle both as Plumian Professor and as director of the Institute of Astronomy at the University of Cambridge. He remembers Hoyle fondly as being always supportive, in spite of the fact that some of Rees's own work on the cosmic microwave background and on quasars helped bring about the collapse of the steady state theory. Rees still holds Hoyle in the highest regard—a photograph of Hoyle hangs on the wall of Rees's office at the Institute of Astronomy. Rees offered two tantalizing potential causes for Hoyle's dissidence. First, he emphasized the negative effects of scientific isolation. He explained that from about the mid-1960s on, Hoyle talked about science almost exclusively with his close collaborators: a very small group that included Jayant Narlikar, Chandra Wickramasinghe, and the Burbidges. Since these scientists rarely if ever disagreed with Hoyle, this was clearly not a good recipe for changing one's views. To my surprise, Rees told me that even though Hoyle had always been very generous and encouraging, he almost never discussed science with him. In fact, Hoyle did not compare notes about new scientific discoveries with any young cosmologists outside his circle of supporters.

Rees made a second interesting observation, which was reminiscent of one of Faulkner's remarks. He noted that in the late stages of their working lives, some scientists lose interest in the routine, incremental advances that normally characterize long stretches of scientific efforts, and they turn their attention to completely new branches of science, sometimes even outside their area of expertise. Rees pointed to Linus Pauling's almost obsessive preoccupation with vitamin C late in his career as an example of this phenomenon, and he held Hoyle's misguided endeavors regarding the origin of life on Earth in a similar light.

There is no doubt that the factors suggested by Rees, Eggleton, and Faulkner played roles in Hoyle's stubbornness. A few statements made by Hoyle himself provide the best evidence. In *Home Is Where the Wind Blows*, he wrote the following striking paragraph:

> The problem with the scientific establishment goes back to the small hunting parties of prehistory. It must then have been the case that, for a hunt to be successful, the entire party was needed. With the direction of prey uncertain, as the direction of the correct theory in science is initially uncertain, the party had to make a decision about which way to go, and then they all had to stick to the decision, even if it was merely made at random. The dissident who argued that the correct direction was precisely opposite from the chosen direction had to be thrown out of the group, just as the scientist today who takes a view different from the consensus finds his papers rejected by journals and his applications for research grants summarily dismissed by state agencies. Life must have been hard in pre-

history, for the more a hunting party found no prey in its chosen direction, the more it had to continue in that direction, for to stop and argue would be to create uncertainty and to risk differences of opinion breaking out, with the group then splitting disastrously apart. This is why the first priority among scientists is not to be correct but for everybody to think the same way. It is this perhaps instinctive primitive motivation that creates the establishment.

One can hardly imagine a stronger advocacy for dissent from mainstream science. Hoyle echoes here the words of the influential second-century physician Galen of Pergamum: "From my very youth I despised the opinion of the multitude and longed for truth and knowledge, believing that there was for man no possession more noble or divine." However, as Rees has pointed out, isolation has its price. Science progresses not in a straight line from A to B but in a zigzag path shaped by critical reevaluation and fault-finding interaction. The continuous evaluation provided by the scientific establishment that Hoyle so despised is what creates the checks and balances that keep scientists from straying too far in the wrong direction. By imposing upon himself academic isolation, Hoyle denied himself these corrective forces.

Hoyle's idiosyncratic ideas on the origin of life had undoubtedly also fueled his refusal to abandon the steady state theory. Here is how Hoyle himself put it:

The proper philosophical point of view, I believe, for thinking about evolution cosmologically involves issues that are superastronomical, as one inevitably gets as soon as one attempts to understand the origin of biological order. Faced with problems of superastronomical order of complexity, biologists have resorted to fairy tales. This is shown by a consideration of the order of the amino acids in any one of hundreds of enzymes [Hoyle estimated that the probability of forming two thousand enzymes *at random* from amino acids was about one in $10^{40,000}$.] . . . to have any hope of solving the problem of biological origins in a rational way *a universe with an essentially unlimited canvas is required* [emphasis added], a universe in which the entropy per unit mass [a measure of the disorder] does not increase inexorably, as it does in big bang cosmologies. It is to provide just such an unlimited canvas that the steady-state theory is required, or so it seems to me.

In other words, Hoyle believed that an evolving universe, with its associated increasing disorder, does not provide the necessary conditions for something as ordered as biology to emerge. He also did not think that the age of the universe, as implied by the value of the Hubble constant, was sufficient for complex molecules to form. I should note that mainstream evolutionary biologists flatly reject this argument. In essence, Hoyle tried to revive the "watchmaker analogy" that characterized all intelligent design arguments by comparing the random origin of a living cell to the likelihood that "a tornado sweeping through a junkyard might assemble a Boeing 747 from the materials therein." Biologist Richard Dawkins labeled this reasoning "Hoyle's fallacy," pointing out that biology does not require intricate life structures to arise in a single step. Organisms that can reproduce themselves are able to generate complexity through successive changes, while inanimate objects are unable to pass on reproductive modifications.

To progress somewhat beyond these partial explanations for Hoyle's blunder, especially when it comes to his apparent denial of having made a mistake, we need to understand the concept of denial a little better. Denial seldom evokes sympathy, especially in scientific circles. Justifiably, scientists regard denial as being contradictory to the research spirit, where old theories have to give way to new ones, when experimental results so require. Research, however, is still carried out by humans, and Sigmund Freud himself had already pos-

tulated that humans have developed denial as one of their defense
mechanisms against traumas or external realities that threaten the
ego. We are all familiar, for instance, with denial as the first of the
five recognized stages of grief. What is perhaps less widely known
is that the experience of being wrong in a major enterprise constitutes such a trauma. The judicial system provides ample evidence
that this is indeed the case. There have been quite a few incidents in
which both victims of violent crimes and prosecutors in such cases
absolutely refused to believe that the person originally found guilty
was actually innocent, even after DNA evidence or new testimony
exonerated this person conclusively. Denial offers the troubled mind
a way to avoid reopening experiences that were thought to have been
brought to a successful closure. To be sure, being wrong in a scientific theory cannot be compared with erring in convicting an innocent person, but the experience is traumatic nonetheless, and we may
assume that denial, in this sense, may have played a part in Hoyle's
blunder.

I have noted several times that the *idea* of a steady state universe
was brilliant at the time it was proposed. In retrospect, the steady
state universe, with its continuous creation of matter, shares many
features with currently fashionable models of an inflationary universe: the conjecture that the cosmos experienced a faster-than-
light growth spurt when it was a fraction of a second old. In some
respects, the steady state universe is simply a universe in which inflation always occurs. Physicist Alan Guth proposed inflation in 1981
to explain, among other things, the cosmic homogeneity and isotropy. Hoyle enjoyed pointing out that in a paper he had published
with Narlikar in 1963, they had shown that their proposed creation field "acts in such a way as to smooth out an initial anisotropy
[dependence on direction] or inhomogeneity [departure from uni-
formity]," and that "it seems that the universe attains the observed
regularity irrespective of initial boundary conditions." These are
precisely the properties now attributed to inflation. Hoyle's brilliance was also revealed in the fact that he belonged to that small
group of scientists capable of investigating two mutually inconsis-

tent theories in parallel. In spite of continuing to hold out against the big bang for his entire life, Hoyle actually contributed important studies to big bang nucleosyntheses, in particular concerning the cosmic helium abundance and the synthesis of elements at very high temperatures.

Lord Rees described Hoyle once as "the most creative and original astrophysicist of his generation." As a humble astrophysicist, I agree wholeheartedly. Hoyle's theories, even when eventually proven wrong, were always dynamizing, and they unfailingly energized entire fields and catalyzed new ideas. It's no wonder that Hoyle's statue (figure 31) now stands just outside the building named after him at the Institute of Astronomy in Cambridge, which he founded in 1966.

Figure 31

As momentous as Hoyle's contributions have been, there is no question that the person who is most responsible for our current understanding of the workings of the cosmos at large is Albert Einstein. His theories of special and general relativity completely revolutionized our perspective on two of the most basic concepts in existence: space and time. Oddly, the phrase "biggest blunder" has become intimately associated with one of the ideas of this most iconic of all scientists.

CHAPTER 10

THE "BIGGEST BLUNDER"

My subject disperses the galaxies, but it unites the earth.
May no "cosmical repulsion" intervene to sunder us!

— SIR ARTHUR EDDINGTON

When I throw my keys up in the air, they reach some maximum height, and then they fall back into my hand. Only for an instant do the keys stay still, as they reach the highest point. Obviously, the gravitational pull of the Earth is responsible for this behavior. If somehow I could propel the keys to a speed exceeding about seven miles per second, they would escape the Earth altogether, as did, for instance, the Pioneer 10 spacecraft, with which communication was lost in 2003, when the probe was at a distance of more than seven billion miles from Earth. However, in the absence of an opposing force, the Earth's gravity alone does not allow for the keys to float suspended in midair.

Two scientists showed independently in the 1920s that the behavior of the cosmic space-time is expected to be very similar. Those two researchers, Russian mathematician and meteorologist Aleksandr Friedmann and Belgian priest and cosmologist Georges Lemaître, applied Einstein's theory of general relativity to the universe as a whole. They soon realized that the gravitational attraction of all the matter and radiation in the universe implies that space-time, Einstein's combination of space and time, can either stretch or contract, but it cannot stably stand still at a fixed extent. These important findings eventually provided the theoretical background for the discovery by Lemaître and Hubble that our universe is expanding. But let's start from the beginning.

In 1917 Einstein himself first attempted to understand the evolution of the entire universe in light of his general relativity equations. This effort initiated the transformation of cosmological problems from speculative philosophy into physics. The expansion of the universe had not been discovered yet. Moreover, not only was Einstein unaware of any observed large-scale motions, but until that time, most astronomers still believed that the universe consisted exclusively of our Milky Way galaxy, with nothing beyond. Astronomer Vesto Slipher's observations of the *redshifts* (the stretchings of light, which were later interpreted as recession velocities of galaxies) of "nebulae" were neither widely known nor understood at the time. Astronomer Heber Curtis did present some preliminary evidence that the Andromeda galaxy, M31, might be outside the Milky Way, but Edwin Hubble confirmed unambiguously this profound fact—that our galaxy is not the entire universe—only in 1924.

Convinced in 1917 that the cosmos was unchanging and static on its largest scales, Einstein had to find a way to keep the universe described by his equations from collapsing under its own weight. To achieve a static configuration with a uniform distribution of matter, Einstein guessed that there had to be some repulsive force that could balance gravity precisely. Consequently, just a little over a year after he had published his theory of general relativity, Einstein came up with what appeared, at least at first glance, to be a brilliant solution. In a seminal paper entitled "Cosmological Considerations on the General Theory of Relativity," he introduced a new term into his equations. This term gave rise to a surprising effect: a repulsive gravitational force! The cosmic repulsion was supposed to act throughout the universe, causing every part of space to be pushing on every other part—just the opposite of what matter and energy do. As we shall soon discover, mass and energy warp space-time in such a way that matter falls together. The fresh cosmological term effectively warped space-time in the opposite sense, causing matter to move apart. The value of a new constant that Einstein introduced (on top of the familiar strength of gravity) determined the strength of the repulsion. The Greek letter lambda, Λ, denoted the new constant, now known as the *cosmological constant*. Einstein demonstrated that he could choose the value of the cosmological constant to precisely balance gravity's attractive and repulsive forces, resulting in a static, eternal, homogeneous, and unchanging universe of a fixed size. This model later became known as "Einstein's universe." Einstein concluded his paper with what turned out to be a pregnant comment: "That term is necessary *only* [*my emphasis*] for the purpose of making possible a quasi-static distribution of matter, as required by the fact of the small velocities of the stars." You'll notice that Einstein talks here about "velocities of stars" and not of galaxies, since the existence and motions of the latter were still beyond the astronomical horizons at the time.

With few exceptions, hindsight is usually 20/20. Cosmologists tend to emphasize the fact that by introducing the cosmological constant, Einstein missed a golden opportunity for a spectacular prediction. Had he stuck with his original equations, he could have predicted more than a decade before Hubble's observations that the universe should be either contracting or expanding. This is certainly true. However, as I shall argue in the next chapter, the introduction of the cosmological constant could have constituted an equally significant prediction.

You may wonder how Einstein could add this new repulsive term into his equations without spoiling general relativity's other successes in explaining several perplexing phenomena. For instance, general relativity elucidated the slight shift in the orbit of the planet Mercury in each successive passage around the Sun. Einstein was, of course, aware that his cosmological constant could undermine agreement with observations, so to avoid undesired consequences, he modified his equations in such a way that the cosmic repulsion increased proportionally to the spatial separation. That is, the repulsion was imperceptible over the distance scales of the solar system, but it became increasingly appreciable over vast cosmological distances. As a result, all the experimental verifications of general relativity (which relied on measurements spanning relatively short distances) could be preserved.

Inexplicably, Einstein did make one surprising mistake in thinking that the cosmological constant would produce a static universe. While the modification did formally allow for a static solution of the equations, that solution described a state of an *unstable equilibrium*—a bit like a pencil standing on its tip or a ball on the top of a hill—the slightest departure from rest resulting in forces moving the system even further away from equilibrium. One can understand this point even without the aid of sophisticated mathematics. The repulsive force increases with distance, while the ordinary attractive force of gravity decreases with distance. Consequently, while one can find a mass density at which the two forces balance each other precisely, any slight perturbation in the form of, say, a small expansion would *increase* the repulsive force and *decrease* the attractive one, resulting in accelerating expansion. Similarly, the slightest contraction would result in total collapse. Eddington was the first to point out this mistake in 1930, and he credited Lemaître with the original perspicacity. However, by then, the fact that the universe was expanding had become widely known, so this particular shortcoming of Einstein's static universe was no longer of any interest. I should also add that in his original paper, Einstein specified neither the physical origin of the cosmological constant nor its precise characteristics. We shall return to these intriguing questions—and, indeed, to the subject of how gravity can exert a repulsive push at all—in the next chapter.

Despite these unresolved issues, Einstein was generally pleased with having succeeded (or so he thought) in constructing a model for a static universe—a cosmos that he regarded as compatible with the prevailing astronomical thinking. Initially, he was also satisfied with the cosmological constant for another reason. The new modification to the original gravitational field equations seemed to attune the theory with some philosophical principles that Einstein had used previously in conceiving general relativity. In particular, the original equations (without the cosmological constant) appeared to require what physicists call "boundary conditions," or specifying a set of values of physical quantities at infinite distances. This was at odds with "the spirit of relativity," in Einstein's words. Unlike Newton's concepts of absolute space and time, one of general relativity's basic premises had been that there is no absolute system of reference. In addition, Einstein insisted that the distribution of matter and energy should determine the structure of space-time. For instance, a universe in which the distribution of matter is trailing off into nothingness would not have been satisfactory, since space-time could not be defined properly without the presence of mass or energy. Yet to Einstein's chagrin, the original equations admitted an *empty* space-time as a solution. He was therefore happy to discover that the static universe turned out not to need any boundary conditions at all, since it was finite and curved on itself like the surface of a sphere, with no boundaries whatsoever. A light ray in this universe came back to its point of origin before starting a new circuit. In this philosophical sense, Einstein, like Plato long before him, always recoiled from the open ended—that which philosopher Georg Wilhelm Hegel referred to as "bad infinity."

I realize that readers who may be a bit rusty on their general relativity would welcome a refresher course, so here is a very brief review of the core principles involved.

Warped Space-Time

In his theory of special relativity, which preceded his articulation of general relativity, Einstein disposed of Newton's notion of an absolute or universal time, one that all clocks would supposedly measure. Newton's goal was to present absolute time and absolute space symmetrically. In that spirit, he stated, "Absolute, true and mathematical time, of itself, and from its own nature, flows equally without relation to anything external." By making the central theme of special relativity the postulate that all observers should measure *the same* speed for light, no matter how fast or in which direction they are moving, Einstein had to pay the price of forever linking space and time together into one interwoven entity called space-time. Numerous experiments have since confirmed the fact that the time intervals measured by two observers moving relative to each other do not agree. Most recently, by comparing two optical atomic clocks connected through an optical fiber, researchers at the National Institute of Standards and Technology managed in 2010 to observe this effect of "time dilation" even for relative speeds as low as twenty-two miles per hour!

Given the central role of light (more generally, electromagnetic radiation) in the theory, special relativity was tailored to agree with the laws that describe electricity and magnetism. Indeed, Einstein entitled his 1905 paper that presented the theory "On the Electrodynamics of Moving Bodies." However, as early as in 1907, he was becoming aware of the fact that special relativity was incompatible with Newton's gravity. Newton's gravitational force was supposed to act instantaneously across all space. The implication was that, for instance, when our Milky Way galaxy and the Andromeda galaxy will collide a few billion years from now, the change in the gravitational field due to the redistribution of mass would be felt simultaneously throughout the entire cosmos. This condition would manifestly conflict with special relativity, since it would mean that information can travel faster than light—impermissible in special relativity. Moreover, the mere concept of worldwide simultaneity would require the existence of the very universal time that special relativity carefully invalidated. While Einstein would not have used this particular example in 1907 because he was unaware of it, he fully understood the principle. To overcome these difficulties—and, in particular, to also allow his theory to apply to accelerated motion— Einstein embarked on a rather winding path that involved many missteps, but one that eventually led him to general relativity.

General relativity is still considered by many to be the most ingenious physical theory ever articulated. The famous physicist Richard Feynman confessed once, "I still can't see how he thought of it." The theory was based largely on two profound insights: (1) the equivalence between gravity and acceleration, and (2) the transformation of the role of space-time from that of a passive spectator to that of a major player in the drama of universal dynamics. First, by contemplating the experience of a person who is free-falling in the gravitational field of the Earth, Einstein realized that acceleration and gravity are essentially indistinguishable. A person living inside a closed elevator on Earth, with the elevator accelerating upward continuously, may think that she lives in a place that has a stronger gravity—a bathroom scale will certainly record a weight that is higher than her normal weight. Similarly, astronauts in the space shuttle were experiencing "weightlessness" because both they and the shuttle were undergoing the same acceleration relative to the Earth. In his Kyoto Lecture in 1922, an impromptu speech to students and faculty members, Einstein described how the idea came to him: "I was sitting in a chair in the patent office in Bern when all of a sudden a thought occurred to me: 'If a person falls freely he will not feel his own weight.' I was startled. This simple thought made a deep impression on me. It impelled me toward a theory of gravitation."

Einstein's second idea turned Newton's gravity on its ear. Gravity is not some mysterious force that acts across space, Einstein contended. Rather, mass and energy warp space-time in the same way that a person standing on a trampoline causes it to sag. Einstein defined gravity as the curvature of space-time. That is, planets move along the shortest paths in the curved space-time created by the Sun, just as a golf ball follows the undulation of the green, or a Jeep negotiates the dunes of the Sahara Desert. Light does not travel in straight lines, either, but curves in the warped neighborhood of large masses. Figure 32 shows a letter written by Einstein in 1913, as he was developing the theory. In the letter, addressed to the American astronomer George Ellery Hale, Einstein explained the bending of light in a gravitational field and the Sun's deflection of light from a distant star. This crucial prediction was first tested in 1919 during an eclipse of the Sun. The person who organized the observations (in Brazil and on Principe Island in the Gulf of Guinea) was Arthur Eddington, and the deviations recorded by his team and by the expedition

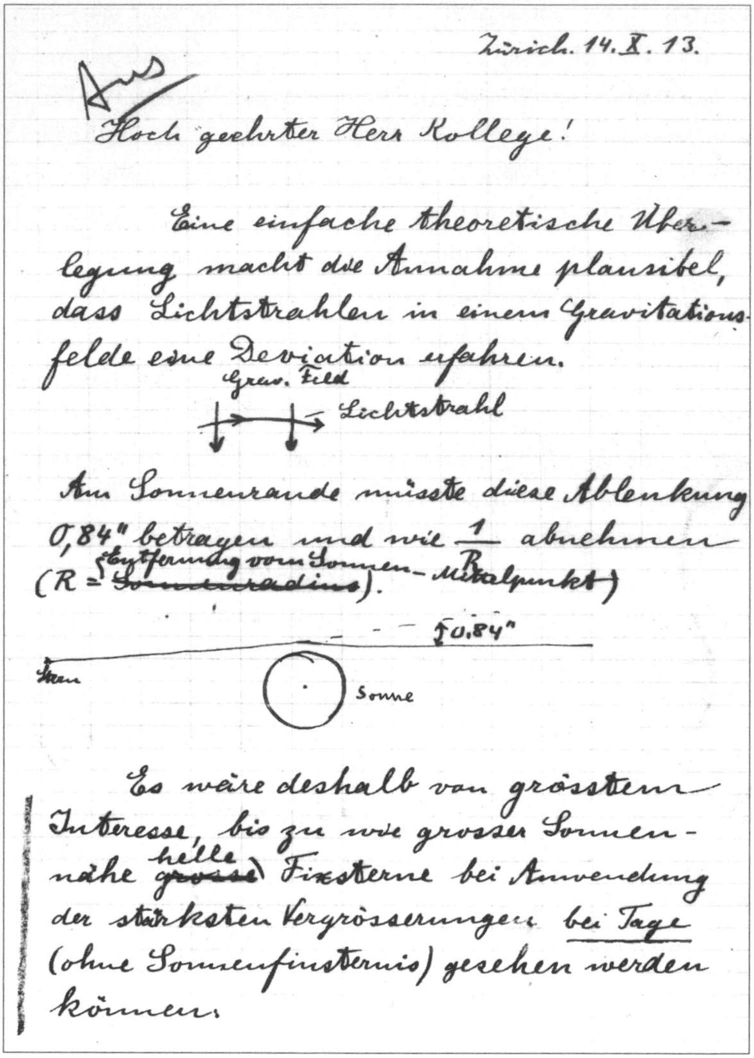

Zürich. 14. X. 13.

Aus

Hoch geehrter Herr Kollege!

Eine einfache theoretische Überlegung macht die Annahme plausibel, dass Lichtstrahlen in einem Gravitationsfelde eine Deviation erfahren.

Grav. Feld

Lichtstrahl

Am Sonnenrande müsste diese Ablenkung 0,84" betragen und wie $\frac{1}{R}$ abnehmen (R = ~~Sonnenradius~~ Entfernung vom Sonnen-Mittelpunkt).

0,84"

Stern

Sonne

Es wäre deshalb von grösstem Interesse, bis zu wie grosser Sonnennähe ~~grosse~~ helle Fixsterne bei Anwendung der stärksten Vergrösserungen bei Tage (ohne Sonnenfinsternis) gesehen werden können.

Figure 32

headed by the Irish astronomer Andrew Crommelin (of about 1.98 and 1.61 seconds of arc) were consistent, within the estimated observational errors, with Einstein's prediction of 1.74 seconds of arc. (Newtonian gravity predicted half that.) Time is "curved" as well in general relativity: Clocks that are near massive bodies tick more slowly than clocks that are far away from them. Experiments have confirmed this effect, which is also taken into account routinely by GPS satellites.

Einstein's pivotal premise in general relativity was a truly revolutionary idea: What we perceive as the force of gravity is merely a manifestation of the fact that mass and energy cause space-time to warp. In this sense, Einstein was closer, at least in spirit, to the geometrical (rather than dynamical) views of the astronomers of ancient Greece than to Newton and his emphasis on forces. Instead of being a rigid and fixed background, space-time can flex, curve, and stretch in response to the presence of matter and energy, and those warps, in turn, cause matter to move the way it does. As the influential physicist John Archibald Wheeler once put it, "Matter tells space-time how to curve, and space-time tells matter how to move." Matter and energy become eternal partners to space and time.

By introducing general relativity, Einstein dazzlingly solved the problem of the faster-than-light propagation of the force of gravity—the predicament that bedeviled Newton's theory. In general relativity, the speed of transmission boils down to how fast ripples in the fabric of space-time can travel from one point to another. Einstein showed that such warps and swells—the geometrical manifestation of gravity—travel precisely at the speed of light. In other words, changes in the gravitation field cannot be transmitted instantaneously.

What's in a Word?

As happy as Einstein might have been with the cosmological constant and his static universe, this satisfaction was soon to evaporate, since new scientific discoveries rendered the concept of a static universe untenable. First, there were a few theoretical disappointments, the earliest of which hit almost immediately. Just one month after the publication of Einstein's cosmological paper, his colleague and friend Willem de Sitter found a solution to Einstein's equations with no matter at all. A cosmos devoid of matter was clearly in contradiction to Einstein's aspiration to connect the geometry of the universe to its mass and energy content. On the other hand, de Sitter himself was quite pleased, since he objected to the introduction of the cosmological constant from day one. In a letter to Einstein dated March 20, 1917, he argued that lambda may have been desirable philosophically but not physically. He was troubled in particular by the fact that he thought that the value of the cosmological constant could not be determined empirically. At that instant, Einstein himself was still keeping an open mind to all options. In his reply to de Sitter, on April 14, 1917, he prophetically wrote a beautiful paragraph, very reminiscent of Darwin's famous "In the distant future . . . light will be thrown on the origin of man" (see chapter 2):

> In any case, one thing stands. The general theory of relativity *allows* the inclusion of $\Lambda g_{\mu\nu}$ [the cosmological term] in the field equations. One day, our actual knowledge of the composition of the fixed star sky, the apparent motions of fixed stars, and the position of spectral lines as a function of distance, will probably have come far enough for us to be able to decide empirically the question of whether or not Λ vanishes. Conviction is a good motive, but a bad judge!

As we shall see in the next chapter, Einstein predicted precisely what astronomers would achieve eighty-one years later. But in 1917, the setbacks just kept coming. Even though de Sitter's model appeared at first blush to be static, that proved to be an illusion. Later work by physicists Felix Klein and Hermann Weyl showed that when test bodies were inserted into it, they were not at rest— rather, they flew away from one another.

The second theoretical blow came from Aleksandr Friedmann. As I noted earlier, Friedmann showed in 1922 that Einstein's equations (with or without the cosmological term) allowed for nonstatic solutions, in which the universe either expanded or contracted. This prompted the disappointed Einstein to write in 1923 to his friend Weyl, "If there is no quasi-static world, then away with the cosmological term." But the most serious challenge was observational. As we have seen in chapter 9, Lemaître (tentatively) and Hubble (unequivocally) showed in the late 1920s that the universe is, in fact, not static—it is expanding. Einstein realized the implications immediately. In an expanding universe, the attractive force of gravity merely slows the expansion. Following Hubble's discovery, therefore, he had to admit that there was no longer a need for an intricate balancing act between attraction and repulsion; consequently, the cosmological constant could be removed from the equations. In a paper published in 1931, he formally abandoned the term, since "the theory of relativity seems to satisfy Hubble's new results more naturally ... without the Λ term." Then, in 1932, in a paper Einstein published together with de Sitter, the authors concluded: "Historically the term containing the 'cosmological constant' Λ was introduced into the field equations in order to enable us to account theoretically for the existence of a finite mean density in a static universe. It now appears that in the dynamical case this end can be reached without the introduction of Λ."

Einstein was aware of the fact that without the cosmological constant, Hubble's measured rate of expansion produced an age for the universe that was uncomfortably short compared with estimated stellar ages, but he was initially of the opinion that the problem might be with the latter. The largest contribution to the error in the observationally determined cosmic expansion rate was corrected only in the 1960s, but uncertainties of a factor of about two in the rate continued to linger until the advent of the Hubble Space Telescope. Surprisingly, however, the banished cosmological constant did return with a bang in 1998.

You'll notice that the language used by Einstein and de Sitter regarding the cosmological constant is benign; they note merely that in an expanding universe, it is not needed. Yet, if you read almost any account of the history of the cosmological constant, you will invariably find the story that Einstein denounced the introduction of this constant into his equations as his "biggest blunder." Did Einstein actually say this, and if so, why?

After scrutinizing all the available documents, I first confirmed something that a few historians of science have already suspected: The tale of Einstein calling the cosmological constant his biggest blunder originated from a single source: the colorful George Gamow. Recall that Gamow was responsible for the idea of big bang nucleosynthesis, as well as for some of the early thinking about the genetic code. James Watson, the codiscoverer of the structure of DNA, once said about Gamow that he was "so very often a step ahead of everybody." Gamow told the "biggest blunder" story in two places. In an article entitled "The Evolutionary Universe," published in the September 1956 issue of *Scientific American,* Gamow wrote, "Einstein remarked to me many years ago that the cosmic repulsion idea was the biggest blunder he had made in his entire life." He repeated the same story [and for some reason, most accounts of the history of the cosmological constant are aware only of this source] in his autobiographical book *My World Line,* which was published posthumously in 1970: "Thus, Einstein's original gravity equation was correct, and changing it [to introduce the cosmological constant] was a mistake. Much later, when I was discussing cosmological problems with Einstein, he remarked that the introduction of the cosmological term was the biggest blunder he ever made in his life."

Since Gamow was known, however, to embellish many of his anecdotes (his first wife said once, "In more than twenty years together, Geo has never been happier than when perpetuating a practical joke"), I decided to dig a bit deeper in an attempt to establish the authenticity of this account. My motivation to investigate this particular quote was enhanced by the fact that the recent resurrection of the cosmological constant has turned "biggest blunder" into one of Einstein's most cited phrases. The last time I checked, there were more than a half million Google pages containing "Einstein" and "biggest blunder"!

I started by trying to ascertain whether Gamow was purporting to actually quote Einstein directly. Unfortunately, each of the two citations presented above appears insufficient, as it stands, to determine whether Gamow was claiming that Einstein himself had used the words "biggest blunder" that he ever made in his life, or whether Gamow was merely reporting the spirit of the conversation. However, in *My World Line,* Gamow continued to say, "But this 'blunder,' rejected by Einstein, is still sometimes used by cosmologists even today, and the cosmological constant denoted by the Greek letter 'Λ' rears its ugly head again and again." The use of quotation marks around the word "blunder" seems at least to suggest that Gamow meant to imply an authentic quote. The fact that Gamow used precisely the same language twice also indicates that he was trying to give the impression, at least, that he was quoting Einstein directly. Note also that Gamow reveals here his own prejudice concerning the cosmological constant, through the phrase "its ugly head."

Intriguingly, I discovered that Einstein did actually use the expression "I made one great mistake in my life," but in an entirely different context. Linus Pauling spoke with Einstein (as one leading scientist and pacifist to another) at Princeton, on November 16, 1954. Immediately following the conversation, Pauling wrote in his diary that Einstein told him the following (figure 33 shows Pauling's diary entry): "He had made one great mistake—when he signed the letter to President Roosevelt recommending that atom bombs be made; but that there was some justification—the danger that the Germans would make them." Clearly, this fact in itself does not necessarily preclude the possibility that Einstein might have used "biggest blunder" also in a scientific context, although the language employed in the conversation with Pauling ("*one* great mistake") does make you wonder.

The second question I wanted to try to settle was that of the circumstances; *when* might Einstein have used this expression with Gamow? In *My World Line* Gamow gives the impression that he and Einstein were very close friends. He describes how during World War II, the two of them served at the same time as consultants in the Division of High Explosives of the US Navy's Bureau of Ordnance. Since Einstein was unable at the time to travel from Princeton to Washington, DC, Gamow recounts, Gamow was "selected," in his words, by the navy to bring documents to Einstein "every other

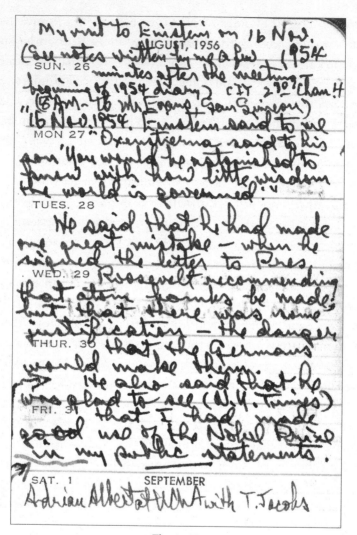

Figure 33

Friday," since he "happened to have known Einstein earlier, on non-military grounds." Gamow goes on to depict a very warm and intimate bond between him and Einstein:

> Einstein would meet me in his study at home, wearing one of his famous soft sweaters, and we would go through all the proposals, one by one . . . After the business part of the

visit was over, we had lunch either at Einstein's home or at the cafeteria of the Institute for Advanced Study, which was not far away, and the conversation would turn to the problems of astrophysics and cosmology . . . I will never forget these visits to Princeton, during which I came to know Einstein much better than I had known him before.

Taking this description as factual, physicist Gino Segrè naturally concluded in his book *Ordinary Geniuses: Max Delbruck, George Gamow, and the Origins of Genomics and Big Bang Cosmology* that Einstein made the "biggest blunder" remark during one of those "World War II Princeton talks." Albrecht Fölsing, who wrote one of the most accurate biographies of Einstein, also assumed that Gamow's account was authentic, and he repeated the alleged "biggest blunder" citation, as did many others. Unfortunately, as I have discovered, the reality was rather different.

Stephen Brunauer was already an accomplished surface scientist when he became, as a lieutenant, head of high explosives research and development for the US Navy during World War II. At one point, he inquired with the army and the civilian divisions whether Einstein was working for them. The answer was negative from both branches. They explained to Brunauer that Einstein was a pacifist, and, furthermore, he was "not interested in anything practical." Unwilling to accept this characterization as definitive, Brunauer visited Einstein at Princeton on May 16, 1943, and he recruited him as a consultant to the navy for a fee of $25 per day. Brunauer was also the officer who recruited Gamow on September 20, 1943. (See his letter to Gamow, figure 34.) In an article published in 1986, entitled "Einstein and the Navy: . . . 'an unbeatable combination,'" Brunauer described the entire episode in detail. He mentioned that in addition to himself, a few other scientists in the division occasionally made use of Einstein's services, including physicists Raymond Seeger, John Bardeen (who went on to win two Nobel Prizes in physics), and George Gamow, as well as chemist Henry Eyring. When explaining Gamow's precise role, Brunauer wrote, "Gamow, in later years, gave

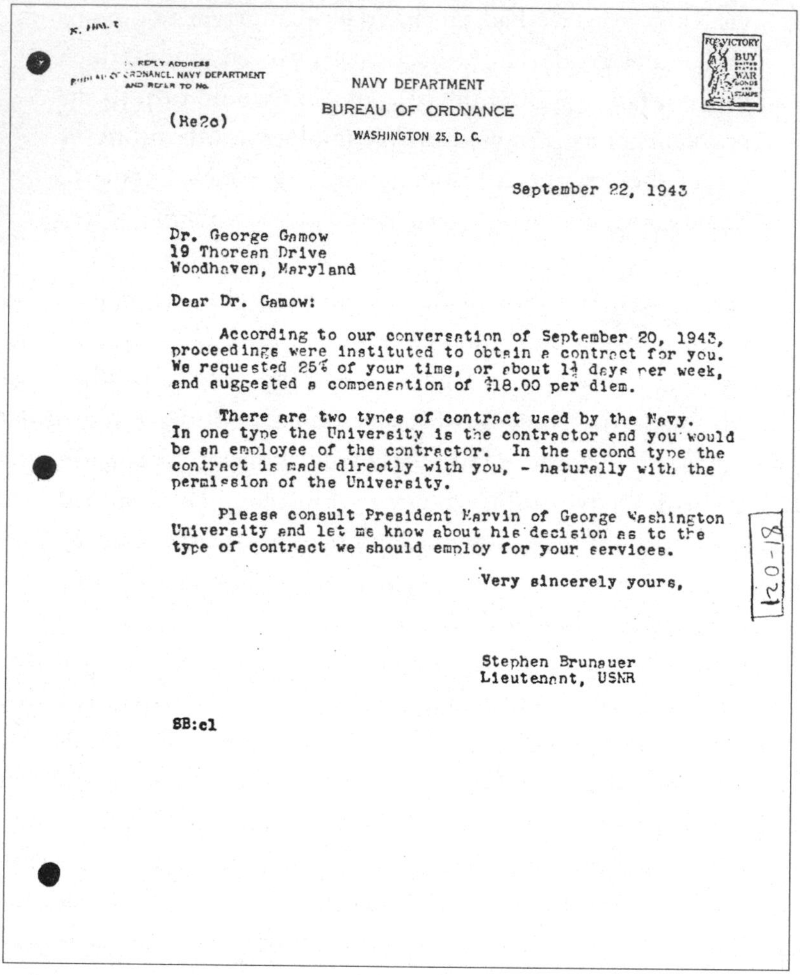

Figure 34

the impression that he was the Navy's liaison man with Einstein, that he visited every two weeks, and the professor 'listened' but made no contribution—all false. The greatest frequency of visits was mine, and that was about every two months."

This narrative clearly sheds a somewhat different light on the Einstein-Gamow interaction. Scrutiny of the few, quite formal letters exchanged between Gamow and Einstein only enhanced my sense that the two men were not close. In one of those, Gamow asked for Einstein's opinion on the idea that the universe as a whole might have nonzero angular momentum (a measure of rotation). To another, Gamow attached his paper on the synthesis of the elements in the big bang. Einstein replied politely to Gamow's letters, but nowhere did he mention the cosmological constant. Perhaps the most telling piece of information in the entire correspondence, however, is a comment Gamow added to Einstein's letter of August 4, 1946. Einstein informed Gamow that he had read the manuscript on big bang nucleosynthesis and that he was "convinced that the abundance of elements as function of atomic weight is a highly important starting point for cosmogonic speculations." Gamow wrote across the bottom of the letter (figure 35), "Of course, the old man agrees with almost anything nowaday."

But if Einstein and Gamow were not close, isn't it surprising that Einstein would use such strong language ("biggest blunder" in his "entire life") concerning the cosmological constant with Gamow, and not with any other of his more intimate friends and colleagues? To explore this point further, I perused Einstein's papers, books, and personal correspondence written later than 1932, for any other mention of the cosmological constant. I used 1932 as the starting point because that was the year in which Einstein and de Sitter declared the cosmological constant unnecessary.

Einstein's writings leave no doubt that following the discovery of the cosmic expansion, he was unhappy with having introduced the cosmological constant in the first place. For instance, in 1942 his assistant and collaborator physicist Peter Bergmann published a book entitled *Introduction to the Theory of Relativity*, which included a foreword by Einstein, who later reviewed the work. The book does not even mention the cosmological constant. However, in the second edition of his own book *The Meaning of Relativity*, Einstein added an appendix in which he did remark on the cosmological term:

> The introduction of the "cosmological member" into the equations of gravity, though possible from the point of view of relativity, is to be rejected from the point of view

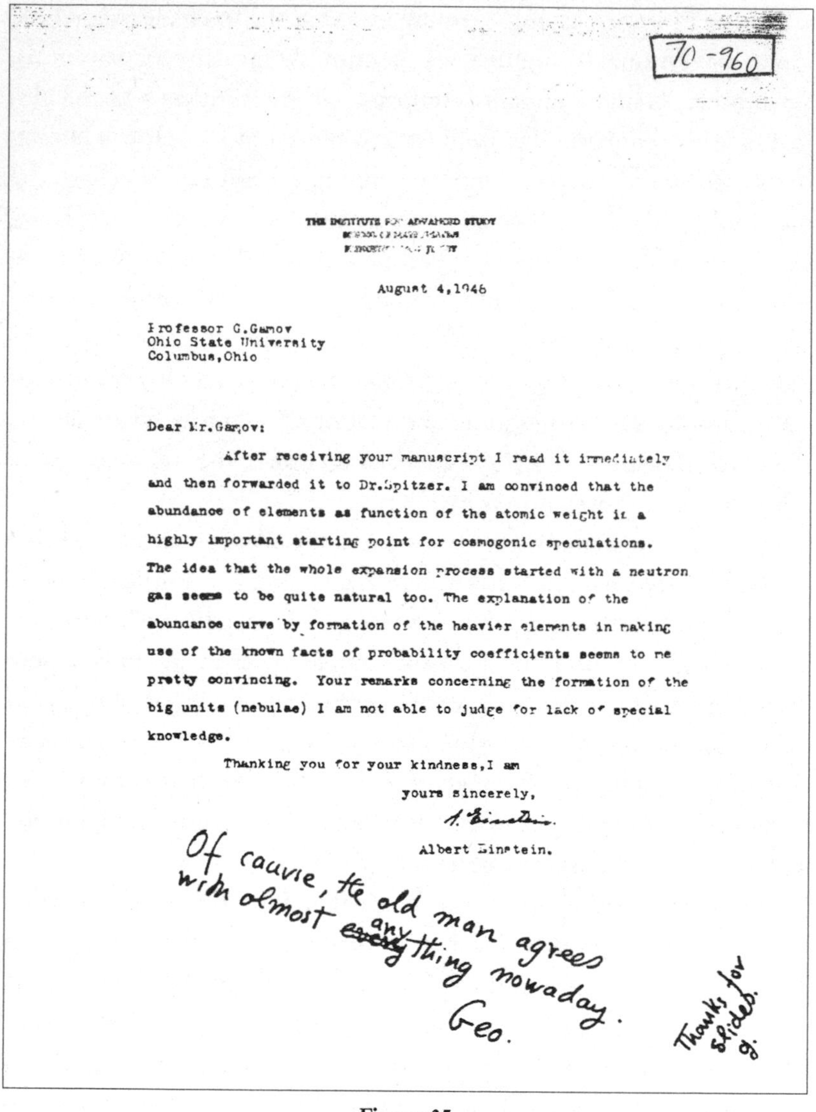

70-960

THE INSTITUTE FOR ADVANCED STUDY

August 4,1946

Professor G.Gamov
Ohio State University
Columbus,Ohio

Dear Mr.Gamov:

After receiving your manuscript I read it immediately and then forwarded it to Dr.Spitzer. I am convinced that the abundance of elements as function of the atomic weight is a highly important starting point for cosmogonic speculations. The idea that the whole expansion process started with a neutron gas seems to be quite natural too. The explanation of the abundance curve by formation of the heavier elements in making use of the known facts of probability coefficients seems to me pretty convincing. Your remarks concerning the formation of the big units (nebulae) I am not able to judge for lack of special knowledge.

Thanking you for your kindness,I am

yours sincerely,

A. Einstein.

Albert Einstein.

Of course, the old man agrees with almost ~~every~~ any thing nowaday.
Geo.

Thanks for slides.
G.

Figure 35

of logical economy. As Friedman[n] was the first to show one can reconcile an everywhere finite density of matter with the original form of the equations of gravity if one admits the time variability of the metric distance of two mass points.

In other words, Einstein recognized that the principles of general relativity allowed the addition of the cosmological repulsion term to the equations, but since it was not needed, he invoked mathematical simplicity to reject it. He then supplemented this comment with a footnote:

> If Hubble's expansion had been discovered at the time of the creation of the general theory of relativity, the cosmological member would never have been introduced. It seems now so much less justified to introduce such a member into the field equations, since its introduction loses its sole original justification—that of lending to a natural solution of the cosmological problem.

In appendix 4 to his popular book *Relativity: The Special and General Theory*, Einstein also noted that the cosmological term "was not required by the theory as such nor did it seem natural from a theoretical point of view." Similarly, the revised, 1958 edition of Nobel laureate Wolfgang Pauli's book *Theory of Relativity* included a supplementary footnote referring to the fact that Einstein was fully aware of the Friedmann and Lemaître solutions and of Hubble's discovery. According to the author, a member of Einstein's inner circle, Einstein subsequently rejected the cosmological term as "superfluous and no longer justified." Pauli commented further that he himself fully accepted Einstein's new standpoint. Nowhere, however, is there any allusion to "biggest blunder."

An analysis of Einstein's entire record about the cosmological constant makes it absolutely clear that he denounced it on two grounds only: an aesthetically motivated simplicity, and the regret over the wrong motivation for its introduction. As I noted in chapter 2, simplicity in terms of the *principles* involved is considered one of the hallmarks of a beautiful theory. To Einstein, simplicity was more than that—it was almost a criterion of reality: "Our experience up to date justifies us in feeling sure that in nature is actualized the ideal of mathematical simplicity." Einstein's experience during the development of general relativity had only enhanced his trust in mathematical principles. When he tried to follow what he thought were the physical constraints, he got nowhere, whereas following the most natural equations from a mathematical perspective opened the door to a "theory of incomparable beauty," in his words. Adding another constant (the cosmological constant) to the equations did not convey reductionistic beauty to Einstein, but he was willing to live with it for as long as it appeared to be imposed by what he perceived as a static reality. Once the cosmos was found to be dynamically expanding, Einstein was delighted to rid his theory of what he now regarded as excess baggage. He articulated his feelings in a letter he wrote to Georges Lemaître on September 26, 1947. This was a reply to a letter that the Belgian cosmologist had sent to Einstein on July 30 of the same year. In that letter (and in an article by Lemaître that followed), Lemaître did his best to persuade Einstein that the cosmological constant was actually necessary to explain a number of cosmic facts, including the age of the universe.

Einstein first admitted that "the introduction of the Λ term offers a possibility" to avoid contradiction with geological ages. Recall that the age of the universe implied by Hubble's original observations was much shorter than the age of the Earth. Lemaître thought that he could resolve this conflict if the equations included the cosmological constant. However, Einstein repeated his reductionist arguments to justify his continuing reluctance to accept the cosmological constant. He wrote:

> Since I have introduced this term I had always a bad conscience. But at that time I could see no other possibility to deal with the fact of the existence of a finite mean density of matter. I found it very ugly indeed that the field law of gravitation should be composed of two logically independent terms which are connected by addition. *About the justification of such feelings concerning logical simplicity it is difficult to argue* [emphasis added]. I cannot help to feel it

strongly and I am unable to believe that such an ugly thing should be realized in nature.

In other words, the original motivation no longer existed, and Einstein felt that aesthetic simplicity was violated, so he did not believe that nature required a cosmological constant. Did he think then that this was his "biggest blunder"? Unlikely. Yes, he was uncomfortable with the concept, saying as early as 1919 that it was "gravely detrimental to the formal beauty of the theory." But general relativity definitely *allowed* for the introduction of the cosmological term, without violating any of the fundamental principles on which the theory had been founded. In this sense, Einstein knew that this was not a blunder at all even before the more recent discoveries concerning the cosmological constant. The experience gained in theoretical physics since Einstein's time has shown that any term allowed by the basic principles is likely to be necessary. Reductionism applies to the fundamentals, not to the specific form of the equations. The laws of physics thus resemble the rules in the Arthurian novel *The Once and Future King* by the English author T. H. White: "Everything that is not forbidden is compulsory."

To conclude, it is virtually impossible to prove beyond any doubt that someone did *not say* something. Still, my best guess, based on the entire body of evidence, is that while Einstein may have had a "bad conscience" about the introduction of the cosmological constant, especially since he missed the chance to predict the cosmic expansion, he never actually called it "the biggest blunder" that he "had ever made." That part was, in my humble opinion, almost certainly Gamow's own hyperbole. Amusingly, in an article entitled "Einstein's Greatest Blunder," University of Manchester astronomer J. P. Leahy commented, "It is just as well that Einstein made his remark to Gamow, otherwise Gamow would have been severely tempted to make it up." My conclusion is that Gamow probably *did* make it up!

You may wonder why this particular quip by Gamow has become one of the most memorable pieces of physics folklore. The answer,

I believe, is threefold. First, people in general, and the media in particular, love superlatives. News in science is always more appealing when it involves "the fastest," "the farthest," "the biggest," or "the first." Einstein, being human, erred many times, but none of his other mistakes created such headlines as his so-called biggest one. Second, Einstein has become the embodiment of genius—the man who purely by his intellectual powers discovered the workings of the universe. He was the scientist who demonstrated that pure mathematics could discover what it creates and also create what it discovers. It has been said about the ancient Greeks that they found the universe a mystery and left it a *polis* (city-state). From the perspective of modern cosmology, this aphorism fits Einstein even better. (Figure 36 shows my favorite picture of Einstein.) The fact that even such a scientific powerhouse is fallible is both fascinating and a wonderful lesson in humility—and in how science truly progresses. Even the most impressive minds are not flawless; they merely pave the way for the next level of understanding. The third reason for the

Figure 36

popularity of the cosmological constant, sometimes called the most famous fudge factor in the history of science, is that it has proven to be the ultimate diehard. Like drug dealer Pablo Escobar and Russian mystic Grigory Rasputin, the cosmological constant has been incredibly hard to kill, even though Einstein denounced it eighty years ago. What's more, not only has this ostensible "blunder" refused to die, but in the past decade it has become the very center of attention. What was it that gave the cosmological constant its nine lives, and why was it thrown into the limelight again?

Hooked on Λ

Even during Einstein's lifetime, there were a few scientists who did not want to give up on the cosmological constant. Physicist Richard C. Tolman, for instance, wrote to Einstein in 1931, "A definite assignment of $\Lambda = 0$, in the absence of experimental determination of its magnitude, seems arbitrary and not necessarily correct." Lemaître, in addition to his general sentiment that Λ should not be rejected only because it had been introduced for the wrong reasons, had two other main motivations for wanting to keep the cosmological constant alive. First, it offered a potential solution to the discrepancy between the perceived young age of the universe (as Hubble's observations seemed to imply) and geological timescales. In some of Lemaître's models, a universe with a cosmological constant could linger for a long time in a coasting state, thus prolonging the age of the cosmos. Lemaître's second reason for championing Λ had to do with his ideas about the formation of galaxies. He conjectured that regions of higher density would be amplified and grow into proto-galaxies during that coasting phase. While this particular idea was shown not to work in the late 1960s, it did help to keep the cosmological constant on the back burner for a while.

Arthur Eddington was another strong supporter of the cosmological constant. So much so, in fact, that at one point he declared defiantly, "Return to the earlier view [without the cosmological constant] is unthinkable. I would as soon think of reverting to Newtonian theory as of dropping the cosmical constant." The main rationale for Eddington's advocacy was that he thought that the repulsive gravity was the true explanation for the observed expansion of the universe. In his words:

> There are only two ways of accounting for large receding velocities of the nebulae: (1) they have been produced by an outward directed force as we have supposed, or (2) as large or larger velocities have existed from the beginning of the present order of things. Several rival explanations of the recession of the nebulae, which do not accept it as evidence of a repulsive force, have been put forward. These necessarily adopt the second alternative, and postulate that the large velocities have existed from the beginning. This might be true; but it can scarcely be called an *explanation* of the large velocities.

In other words, Eddington recognized that even without the cosmological constant, general relativity allowed for an expanding universe solution. However, this solution had to assume that the cosmos started with large velocities, without providing an explanation for those particular initial conditions. The *inflationary model*—the idea that the universe underwent a stupendous expansion when it was only a fraction of a second old—was born out of a similar dissatisfaction with having to rely on specific initial conditions as a cause for observed cosmic effects. For instance, inflation is assumed to have puffed up the universe's fabric so much that it flattened the cosmic geometry. At the same time, inflation is believed to have been the agent that took quantum fluctuations of subatomic size in the density of matter and inflated them to cosmological scales. These were the density enhancements that later became the seeds for the formation of cosmic structure.

As I have noted already in chapter 9, Hoyle's steady state model of 1948 did reproduce some of the features of inflationary cosmology. The field term that Hoyle had introduced into Einstein's equations for the continuous creation of matter acted in many ways like a cosmological constant. In particular, it caused the universe to expand exponentially. Consequently, steady state cosmology helped keep some form of the cosmological repulsive factor in vogue for another fifteen years or so.

When astronomer and longtime Hoyle supporter William McCrea came to summarize the then-prevailing ideas about the cosmological constant in 1971, he distinguished presciently between two possibilities: Either general relativity is a complete, self-consistent theory, or general relativity should be regarded only as one part of a more comprehensive "theory of everything" that describes the cosmos and all phenomena within it. In the first case, McCrea noted, the cosmological constant becomes a nuisance, since its value cannot be determined from within the theory itself. In the second, he argued insightfully, the value of the cosmological constant may be fixed through the connection between general relativity and other relevant branches of physics. As we shall soon see, physicists are trying to understand the nature of the cosmological constant precisely through their efforts to unify the large and the small—general relativity with quantum mechanics.

CHAPTER 11

OUT OF EMPTY SPACE

If we admit that ether is to some degree condensable and extensible, and believe that it extends through all space, then we must conclude that there is no mutual gravitation between its parts, and cannot believe that it is gravitationally attracted by the sun or the earth or any ponderable matter; that is to say, we must believe ether to be a substance outside the law of universal gravitation.

— LORD KELVIN

The cosmological constant introduced into the physics vocabulary a repulsive gravitational force that is proportional to distance and acts over and above the ordinary gravitational attraction between masses. As with so many other physical concepts, Newton was the first to consider the effects of a similar force. In his celebrated *Principia,* he discussed, in addition to the normal force of gravity, a force that "increases in a simple ratio of the distance." Newton was able to show that for this type of force, as with gravity, one could treat spherical masses as if all the mass was concentrated at their centers. What he did not do, however, was to fully examine the problem for the case in which the two forces act in tandem. Newton might have paid more attention to this scenario had he realized, or taken more seriously, the fact that his law of gravitation could not easily be applied to the universe as a whole. If one attempts to calculate the gravitational force at any point in a cosmos of infinite extent and uniform density, the computation does not yield any definite value. The situation is a bit like trying to calculate the sum of the infinite sequence 1–1+1–1+1–1 . . . The result depends on where you stop.

Toward the end of the nineteenth century, a few physicists attempted to find a way out of this conundrum. They suggested solutions ranging from small modifications to Newton's law of gravitation, to the introduction of more exotic concepts such as negative masses. The ubiquitous Lord Kelvin proposed, for instance, that the ether—the stuff then presumed to permeate all space—does not gravitate at all. (See his quote at the beginning of this chapter.) Eventually, all these early endeavors culminated in Einstein's theory of general relativity and the subsequent augmentation of its equations by the cosmological constant. As we have seen, however, Einstein repudiated this term later, and except for its short-lived reincarnation as part of Hoyle's steady state cosmology, it was essentially banished from the theory for a few decades. Astronomical observations of the late 1960s provided the impetus for the next rise of this phoenix from its ashes. Astronomers seemed to find an excess in the counts of quasars clustered around an epoch of about ten billion years ago. This overdensity could be explained if the size of the universe somehow lingered for a while around the dimensions it had at that time— about one-third of its current extent. Indeed, a few astrophysicists showed that such cosmic loitering could be obtained in Lemaître's model, since that involved (through its employment of the cosmological constant) a leisurely coasting, quasi-static phase. Even though this particular model did not survive for long, it did draw attention to one potential interpretation of the cosmological constant: that of the *energy density of empty space*. This idea is so fundamental, and yet so mind boggling that it deserves some explanation.

From the Largest to the Smallest Scales

By definition, mathematical equations are expressions or propositions asserting the equality of two quantities. Einstein's most famous equation, $E = mc^2$, for instance, expresses the fact that the energy associated with a given mass (on the left-hand side of the equality sign) is equal to the product of that mass and the square of the speed of light (on the right-hand side). Einstein's original equation of general relativity was of the following form: It had on its left-hand side a term describing the curvature of space, and on the right-hand side a term specifying the distribution of mass and energy (multiplied by Newton's constant denoting the strength of the gravitational force). This was a clear manifestation of the essence of general relativity: Matter and energy (right-hand side) determine the geometry of space-time (left-hand side), which is the expression of gravity. When he introduced the cosmological constant, Einstein added it on the left-hand side (multiplied by a quantity that defines distances), since he thought of it as yet another geometrical property of space-time. However, if one moves this term to the right-hand side, it acquires a whole new physical meaning. Instead of describing the geometry, the cosmological term is now part of the cosmic energy budget. The characteristics of this new form of energy, however, are different from those of the energy associated with matter and radiation in two important ways. First, while the density of matter (both ordinary and the one called "dark," which does not emit light) decreases as the universe expands, the density of the energy corresponding to the cosmological constant remains eternally *constant*. And if that is not strange enough, this new form of energy has a negative pressure!

Negative pressure sucks. This is not a joke; positive pressure, like that exerted by a compressed regular gas, pushes outward. Negative pressure, on the other hand, sucks inward instead of pushing outward. This property turns out to be crucial, since in general relativity, in addition to mass and energy, pressure is also a source of gravity—it applies its own gravitational force. Moreover, whereas positive pressure generates an attractive force of gravity, negative pressure contributes a *repulsive gravitational force* (a feature that probably makes Newton turn in his grave). This was precisely the attribute of the cosmological constant that Einstein had used in his attempt to keep the universe static. The basic symmetry of general relativity, that the laws of nature should make the same predictions in different frames of reference, implies that only the vacuum—literally empty space—can have an energy density that does not dilute upon expansion. Indeed, how can empty space dilute any further? But energy of the vacuum? Why does empty space have any energy at all? Isn't empty space simply "nothing"?

Not in the weird world of quantum mechanics. When one enters the subatomic realm, the vacuum is far from being nothing. In fact, it is a frenzy of virtual (in the sense that they cannot be observed directly) pairs of particles and antiparticles that pop in and out of existence on fleetingly short timescales. Consequently, even empty space can be endowed with an energy density and, concomitantly, can be a source of gravity. This is an entirely different *physical* interpretation from the one originally suggested by Einstein. Einstein regarded his cosmological constant as a potential peculiarity of space-time—describing the universe on its cosmic largest scales. The identification of the cosmological constant with the energy of empty space, even though mathematically equivalent, intimately relates it to the smallest subatomic scales—the province of quantum mechanics. McCrea's observation in 1971 that one could perhaps determine the value of the cosmological constant from physics outside classical general relativity proved to be truly visionary.

I should note that Einstein himself made one interesting attempt to connect the cosmological constant to elementary particles. In what could be regarded as his first foray into the arena of trying to unify gravity and electromagnetism, Einstein proposed in 1919 that perhaps electrically charged particles are being held together by gravitational forces. This led him to an electromagnetic constraint on the value of the cosmological constant. Apart, however, from one additional short note on the subject in 1927, Einstein never returned to this topic.

The idea that the vacuum is not empty but, rather, could contain a vast amount of energy is not really new. It was first proposed by the German physical chemist Walther Nernst in 1916, but since he was interested primarily in chemistry, Nernst did not consider the implications of his idea for cosmology. The practitioners of quantum mechanics in the 1920s, Wolfgang Pauli in particular, did discuss the fact that in the quantum domain the lowest possible energy of any field is not zero. This so-called *zero-point energy* is a consequence of the wavelike nature of quantum mechanical systems, which causes them to undergo jittery fluctuations even in their ground state. However, even Pauli's conclusions did not propagate into cosmological considerations. The first person to specifically connect the cosmological constant to the energy of empty space was Lemaître. In a paper published in 1934, not long after he had met with Einstein, Lemaître wrote, "Everything happens as though the energy *in vacuo* would be different from zero." He then went on to say that the energy density of the vacuum must be associated with a negative pressure, and that "this is essentially the meaning of the cosmological constant Λ." Figure 37 shows Einstein and Lemaître meeting in Pasadena in January 1933.

As perceptive as Lemaître's comments were, the subject lay dormant for more than three decades until a brief revival of interest

Figure 37

in the cosmological constant attracted the attention of the versatile Jewish Belarusian physicist Yakov Zeldovich. In 1967 Zeldovich made the first genuine attempt to calculate the contribution of vacuum jitters to the value of the cosmological constant. Unfortunately, along the way, he made some ad hoc assumptions without articulating his reasoning. In particular, Zeldovich assumed that most of the zero-point energies somehow cancel out, leaving only the gravitational interaction between the virtual particles in the vacuum. Even with this unjustified omission, the value that he obtained was totally unacceptable; it was about a billion times larger than the energy density of all the matter and radiation in the observable universe.

More recent attempts to estimate the energy of empty space have only exacerbated the problem, producing values that are much, much higher—so high, in fact, that they cannot be considered anything but absurd. For instance, physicists first assumed naïvely that they could sum up the zero-point energies up to the scale where our theory of gravity breaks down. That is, to that point at which the universe was so small that one needs to have a quantum theory of gravity (a theory that does not exist currently). In other words, the hypothesis was that the cosmological constant should correspond to the cosmic density when the universe was only a tiny fraction of a second old, even before the masses of the subatomic particles were imprinted. However, when particle physicists carried out that estimate, it resulted in a value that was about 123 orders of magnitude (1 followed by 123 zeros) greater than the combined cosmic energy density in matter and radiation. The ludicrous discrepancy prompted physics Nobel laureate Steven Weinberg to dub it "the worst failure of an order-of-magnitude estimate in the history of science." Obviously, if the energy density of empty space were truly that high, not only would galaxies and stars not have existed but also the enormous repulsion would have instantly torn apart even atoms and nuclei. In a desperate attempt to correct the guesstimate, physicists used symmetry principles to conjecture that adding up the zero-point energies should be cut off at some lower energy. Dismally, even though the revised estimate resulted in a considerably lower value, the energy was still some 53 orders of magnitude too high.

Faced with this crisis, some physicists resorted to believing that a yet-undiscovered mechanism somehow completely cancels out all the different contributions to the energy of the vacuum, to produce a value of exactly zero for the cosmological constant. You'll recognize that mathematically speaking, this is precisely equivalent to Einstein's simple removal of the cosmological constant from his equations. Assuming that the cosmological constant vanishes means that the repulsive term need not be included in the equation. The reasoning, however, was completely different. Hubble's discovery of the cosmic expansion quickly subverted Einstein's original motivation for introducing the cosmological constant. Even so, many physicists regarded as unjustified the assignment of the specific value of zero to lambda for the mere sake of brevity or as remedy to a "bad conscience." In its modern guise as the energy of empty space, on the other hand, the cosmological constant appears to be obligatory from the perspective of quantum mechanics, unless all the different quantum fluctuations somehow conspire to add up to zero. This inconclusive, frustrating situation lasted until 1998, when new astronomical observations turned the entire subject into what is arguably the most challenging problem facing physics today.

The Accelerating Universe

Since Hubble's observations in the late 1920s, we knew that we live in an expanding universe. Einstein's theory of general relativity provided the natural interpretation of Hubble's findings: The expansion is a stretching of the fabric of space-time itself. The distance between any two galaxies increases just as the distance between any two paper chads glued to the surface of a spherical balloon would increase if the balloon were inflated. However, in the same way that the Earth's gravity slows down the motion of any object thrown upward, one would anticipate that the cosmic expansion should be slowing, due to the mutual gravitational attraction of all the matter

and energy within the universe. But in 1998 two teams of astronomers, working independently, discovered that the cosmic expansion is not slowing down; in fact, over the past six billion years, it
has been speeding up! One team, the Supernova Cosmology Project, was led by Saul Perlmutter of the Lawrence Berkeley National
Laboratory, and the other, the High-Z Supernova Search Team, was
led by Brian Schmidt of Mount Stromlo and Siding Spring Observatory and Adam Riess of the Space Telescope Science Institute and the
Johns Hopkins University.

The discovery of accelerating expansion came as a shock initially, since it implied that some form of repulsive force—of the type
expected from the cosmological constant—propels the universe's
expansion to speed up. To reach their surprising conclusion, the
astronomers relied on observations of very bright stellar explosions
known as Type 1a supernovae. These exploding stars are so luminous
(at maximum light, they may outshine their entire host galaxies) that
they can be detected (and the evolution of their brightness followed)
more than halfway across the observable universe. In addition, what
makes Type 1a supernovae particularly suitable for this type of study
is the fact that they are excellent *standard candles:* Their intrinsic
luminosities at peak light are nearly the same, and the small deviations from uniformity that exist can be calibrated empirically. Since
the observed brightness of a light source is inversely proportional to
the square of its distance—an object that is three times farther than
another is nine times dimmer—knowledge of the intrinsic luminosity combined with measurement of the apparent one allows for a
reliable determination of the source's distance.

Type 1a supernovae are very rare, occurring roughly only once
per century in a given galaxy. Consequently, each team had to
examine thousands of galaxies to collect a sample of a few dozen
supernovae. The astronomers determined the distances to these
supernovae and their host galaxies, and the recession velocities of
the latter. With these data at hand, they compared their results with
the predictions of a linear Hubble's law. If the expansion of the universe were indeed slowing, as everyone expected, they should have

found that galaxies that are, say, two billion light-years away, appear brighter than anticipated, since they would be somewhat closer than where uniform expansion would predict. Instead, Riess, Schmidt, Perlmutter, and their colleagues found that the distant galaxies appeared *dimmer* than expected, indicating that they had reached a larger distance. A precise analysis showed that the results imply a cosmic acceleration for the past six billion years or so. Perlmutter, Schmidt, and Riess shared the 2011 Nobel Prize in physics for their dramatic discovery.

Since the initial discovery in 1998, more pieces of this puzzle have emerged, all corroborating the fact that some new form of a smoothly distributed energy is producing a repulsive gravity that is pushing the universe to accelerate. First, the sample of supernovae has increased significantly and now covers a wide range of distances, putting the findings on a much firmer basis. Second, Riess and his collaborators have shown by subsequent observations that an earlier epoch of deceleration preceded the current six-billion-year-long accelerating phase in the cosmic evolution. A beautifully compelling picture emerges: When the universe was smaller and much denser, gravity had the upper hand and was slowing the expansion. Recall, however, that the cosmological constant, as its name implies, does not dilute; the energy density of the vacuum is constant. The densities of matter and radiation, on the other hand, were enormously high in the very early universe, but they have decreased continuously as the universe expanded. Once the energy density of matter dropped below that of the vacuum (about six billion years ago), acceleration ensued.

The most convincing evidence for the accelerating universe came from combining detailed observations of the fluctuations in the cosmic microwave background by the Wilkinson Microwave Anisotropy Probe (WMAP) with those of supernovae, and supplementing those observations with separate measurements of the current expansion rate (the Hubble constant). Putting all of the observational constraints together, astronomers were able to determine precisely the current contribution of the putative vacuum energy to the total cosmic energy budget. The observations revealed that matter (ordinary and dark together) contributes only about 27 percent of the universe's energy density, while "dark energy"—the name given to the smooth component that is consistent with being the vacuum energy—contributes about 73 percent. In other words, Einstein's diehard cosmological constant, or something very much like its contemporary "flavor"—the energy of empty space—is currently the dominant energy form in the universe!

To be clear, the measured value of the energy density associated with the cosmological constant is still some 53 to 123 orders of magnitude smaller than what naïve calculations of the energy of the vacuum produce, but the fact that it is definitely not zero has frustrated much wishful thinking on the part of many theoretical physicists. Recall that given the incredible discordance between any reasonable value for the cosmological constant—one that the universe could accommodate without bursting at the seams—and the theoretical expectations, physicists were anticipating that some yet-undiscovered symmetry would lead to the complete cancellation of the cosmological constant. That is, they hoped that the different contributions of the various zero-point energies, as large as they might be individually, would come in pairs of opposite signs so that the net result would be zero.

Some of these expectations were hung on concepts such as supersymmetry: particle physicists predict that every particle we know and love, such as electrons and quarks (the constituents of protons and neutrons), should have yet-to-be-found supersymmetric partners that have the same charges (for example, electrical and nuclear), but spins removed by a half quantum mechanical unit. For instance, the electron has a spin of ½, and its "shadow" supersymmetric partner is supposed to have spin of 0. If all superpartners were also to have the same mass as their known partners, then the theory predicts that the contribution of each such pair would indeed cancel out. Unfortunately, we know that the superpartners of the electron, the quark, and the elusive neutrino cannot have the same mass, respectively, as the electron, quark, and neutrino, or they would have been discovered already. When this fact is taken into account, the total contribution to the vacuum energy is larger than the observed one by some 53 orders of magnitude. One might still have hoped that another, yet-unthought-of symmetry would produce the desired cancellation. However, the breakthrough measurement of the cosmic acceleration has shown that this is not very likely. The exceedingly small but nonzero value of the cosmological constant has convinced many theorists that it is hopeless to seek an explanation relying on symmetry arguments. After all, how can you reduce a number to 0.0000000000000000000000000000000000000000000000000000001 of its original value without canceling it out altogether? This remedy seems to require a level of fine-tuning that most physicists are unwilling to accept. It would have been much easier, in principle, to imagine a hypothetical scenario that would make the vacuum energy precisely zero than one that would set it to the observed minuscule value. So, is there a way out? In desperation, some physicists have taken to relying on one of the most controversial concepts in the history of science—anthropic reasoning—a line of thought in which the mere existence of human observers is assumed to be part of the explanation. Einstein himself had nothing to do with this development, but it was the cosmological constant—Einstein's brainchild or "blunder"—that has convinced quite a few of today's leading theorists to consider this condition seriously. Here is a concise explanation of what the fuss is all about.

Anthropic Reasoning

Almost everybody would agree that the question "Does extraterrestrial intelligent life exist?" is one of the most intriguing questions in science today. That this is a reasonable question to ask stems from an important truth: The properties of our universe, and the laws governing it, have allowed complex life to emerge. Obviously, the precise biological peculiarities of humans depend crucially on the Earth's properties and its history, but some basic requirements would seem necessary for any form of intelligent life to materialize. For instance, galaxies composed of stars, and planets orbiting at least some of those stars, appear to be reasonably generic. Similarly, nucleosynthesis in stellar interiors had to forge the building blocks of life: atoms such as carbon, oxygen, and iron. The universe also had to provide for a sufficiently hospitable environment—for a long enough time—that these atoms could combine and form the complex molecules of life, enabling primitive life to evolve to its "intelligent" phase.

In principle, one could imagine "counterfactual" universes that are not conducive for the appearance of complexity. For instance, consider a universe harboring the same laws of nature as ours, and the same values of all the "constants of nature" but one. That is, the strengths of the gravitational, electromagnetic, and nuclear forces are identical to those in our universe, as are the ratios of the masses of all the elementary particles. However, the value of one parameter— the cosmological constant—is a thousand times higher in this hypothetical universe. In such a universe, the repulsive force associated with the cosmological constant would have resulted in such a rapid expansion that no galaxies could have ever formed.

As we have seen, the question we have inherited from Einstein was this: Why should there be a cosmological constant at all? Modern physics transformed that question into: Why should empty space exert a repulsive force? However, owing to the discovery of accelerating expansion, we now ask: Why is the cosmological constant (or the force exerted by the vacuum) so small? In 1987, in the wake of all the previous failed attempts to put a cap on the energy of empty space, physicist Steven Weinberg came up with a bold "What if?" question. What if the cosmological constant is not truly fundamental—explicable within the framework of a "theory of everything"— but *accidental*? That is, imagine that there exists a vast ensemble of universes—a "multiverse"—and that the cosmological constant may assume different values in different universes. Some universes, such as the counterfactual one we discussed with a thousandfold larger lambda, would not have developed complexity and life. We humans find ourselves in one of those universes that are "biophilic." In such a case, no grand unified theory of the basic forces would fix the value of the cosmological constant. Rather, the value would be determined by the simple requirement that it should fall within the range that would allow humans to evolve. In a universe with too large a cosmological constant, there would be no one to ask the question about its value. Physicist Brandon Carter, who first presented this type of argument in the 1970s, dubbed it the "anthropic principle." The attempts to delineate the "pro-life" domains are accordingly described as anthropic reasoning. Under what conditions can we even attempt to apply this type of reasoning to explain the value of the cosmological constant?

In order to make any sense at all, anthropic reasoning has to rely on three basic assumptions:

1. Observations are subjected to a "selection bias"—filtering of physical reality—even merely by the fact that they are executed by humans.
2. Some of the nominal "constants of nature" are accidental rather than fundamental.
3. Our universe is but one member of a gigantic ensemble of universes.

Let me examine very briefly each one of these points and attempt to assess its viability.

Statisticians always dread selection biases. These are distortions of the results, introduced either by the data-collecting tools or by the method of data accumulation. Here are a few simple examples to demonstrate the effect. Imagine that you want to test an investment strategy by examining the performance of a large group of stocks against twenty years' worth of data. You might be tempted to include in the study only stocks for which you have complete information over the entire twenty-year period. However, eliminating stocks that stopped trading during this period would produce biased results, since these were precisely the stocks that did not survive the market.

During World War II, the Jewish Austro-Hungarian mathematician Abraham Wald demonstrated a remarkable understanding of selection bias. Wald was asked to examine data on the location of enemy fire hits on bodies of returning aircraft, to recommend which parts of the airplanes should be reinforced to improve survivability. To his superiors' amazement, Wald recommended adding armor to the locations that showed *no* damage. His unique insight was that the bullet holes that he saw in surviving aircraft indicated places where an airplane could be hit and still endure. He therefore concluded that the planes that had been shot down were probably hit precisely in those places where the persevering planes were lucky enough not to have been hit.

Astronomers are very familiar with the *Malmquist bias* (named after the Swedish astronomer Gunnar Malmquist, who greatly elaborated upon it in the 1920s). When astronomers survey stars or galaxies, their telescopes are sensitive only down to a certain brightness. However, objects that are intrinsically more luminous can be observed to greater distances. This will create a false trend of increasing average intrinsic brightness with distance, simply because the fainter objects will not be seen.

Brandon Carter pointed out that we shouldn't take the Copernican principle—the fact that we are nothing special in the cosmos—too far. He reminded astronomers that humans are the ones who make observations of the universe; consequently, they should not be too surprised to discover that the properties of the cosmos are consistent with human existence. For instance, we could not discover that our universe contains no carbon, since we are carbon-based lifeforms. Initially, most researchers took Carter's anthropic reasoning to be nothing more than a trivially obvious statement. Over the past couple of decades, however, the anthropic principle has gained some popularity. Today quite a few leading theorists accept the fact that in the context of a multiverse, anthropic reasoning can lead to a natural explanation for the otherwise perplexing value of the cosmological constant. To recapitulate the argument, if lambda were much larger (as some probabilistic considerations seem to require), then the cosmic acceleration would have overwhelmed gravity before galaxies had a chance to form. The fact that we find ourselves here in the Milky Way galaxy necessarily biases our observations to low values of the cosmological constant in our universe.

But how reasonable is the assumption that some physical constants are "accidental"? A historical example can help clarify the concept. In 1597 the great German astronomer Johannes Kepler published a treatise known as *Mysterium Cosmographicum* (*The Cosmic Mystery*). In this book, Kepler thought that he had found the solution to two bewildering cosmic enigmas: Why were there precisely six planets in the solar system (only six were known at this time) and what determined the sizes of the planetary orbits? Even in Kepler's time, his answers to these riddles were borderline crazy. He constructed a model for the solar system by embedding the five regular solids known as the *Platonic solids* (tetrahedron, cube, octahedron, dodecahedron, and icosahedron) inside each other. Together with an outer sphere corresponding to the fixed stars, the solids determined precisely six spacings, which to Kepler "explained" the number of the planets. By choosing a particular order for which solid to embed in which, Kepler was able to achieve approximately the correct relative sizes for the orbits in the solar system. However, the main problem with Kepler's model was not in its geometrical details—after all, Kepler used the mathematics that he knew to explain existing observations. The key failure was that Kepler did not realize that neither the number of planets nor the sizes of their orbits were fundamental quantities—ones that can be explained from first principles. While the laws of physics indeed govern the general process of planet formation from a protoplanetary disk of gas and dust, the particular environment of any young stellar object determines the end result.

We now know that there are billions of extrasolar planets in the Milky Way, and each planetary system is different in terms of its members and orbital properties. Both the number of the planets and the dimensions of their circuits are accidental, as is, for instance, the precise shape of any individual snowflake.

There is one particular quantity in the solar system that has been crucial for our existence: the distance between the Earth and the Sun. The Earth is in the Sun's habitable zone—the narrow circumstellar band that allows for liquid water to exist on the planet's surface. At much closer distances, water evaporates, and at much larger ones, it freezes. Water was essential for life to emerge on Earth, since molecules could combine easily in the young Earth's "soup" and could form long chains while being sheltered from harmful ultraviolet radiation. Kepler was obsessed with the idea of finding a first-principles explanation to the Earth-Sun distance, but this obsession was misguided. There was nothing to prevent the Earth (in principle) from forming at a different distance. But had that distance been significantly larger or smaller, there would have been no Kepler to wonder about it. Among the billions of solar systems in the Milky Way galaxy, many probably do not harbor life, since they don't have the right planet in the habitable zone around the host star. Even though the laws of physics did determine the orbit of the Earth, there is no deeper explanation for its radius other than the fact that had it been very different, we wouldn't be here.

This brings us to the last necessary ingredient of anthropic reasoning: For the explanation of the value of the cosmological constant in terms of an accidental quantity in a multiverse to hold any water, there must be a multiverse. Is there? We don't know, but that has never stopped smart physicists from speculating. What we do know is that in one theoretical scenario known as "eternal inflation," the dramatic stretching of space-time can produce an infinite and everlasting multiverse. This multiverse is supposed to continually generate inflating regions, which evolve into separate "pocket universes." The big bang from which our own "pocket universe" came into existence is just one event in a much grander scheme of an exponentially expanding substratum. Some versions of "string theory" (now sometimes called "M-theory") also allow for a huge variety of universes (more than 10^{500}!), each potentially characterized by different values of physical constants. If this speculative scenario is correct, then what we have traditionally called "the universe" could indeed be just one piece of space-time in a vast cosmic landscape.

One should not get the impression that all (or even most) physicists believe that the solution to the puzzle of the energy of empty space will come from anthropic reasoning. The mere mention of the "multiverse" and "anthropics" tends to raise the blood pressure of some physicists. There are two main reasons for this adverse reaction. First, as already mentioned in chapter 9, ever since the seminal work of philosopher of science Karl Popper, for a scientific theory to be worthy of its name, it has to be falsifiable by experiments or observations. This requirement has become the foundation of the "scientific method." An assumption about the existence of an ensemble of potentially unobservable universes appears, at first glance at least, to be in conflict with this prerequisite and therefore in the realm of metaphysics rather than physics. Note, however, that the boundary between what we define as observable and what is not is unclear. Consider, for instance, the "particle horizon": that surface around us from which radiation emitted at the big bang is just reaching us. In the Einstein–de Sitter model—the model for a homogeneous, isotropic, constant curvature universe, with no cosmological constant— the cosmic expansion decelerates, and one could safely expect that all the objects currently lying beyond the horizon will eventually become observable in the distant future. But since 1998, we know that we don't live in an Einstein–de Sitter cosmos: our universe is accelerating. In this universe any object now beyond the horizon will stay beyond the horizon forever. Moreover, if the accelerating expansion continues, as anticipated from a cosmological constant, even galaxies that we can now see will become invisible to us! As their recession speed approaches the speed of light, their radiation will stretch (redshift) to the point where its wavelength will exceed the size of the universe. (There is no limit on how fast space-time can stretch, since no mass is really moving.) So even our own accelerating universe contains objects that neither we nor future generations of astronomers will ever be able to observe. Yet we would not consider such objects as belonging to metaphysics. What could then give us confidence in potentially unobservable universes? The answer is a natural extension of the scientific method: We can believe in their existence if they are predicted by a theory that gains credibility because it is corroborated in other ways. We believe in the properties of black holes because their existence is predicted by general relativity—a theory that has been tested in numerous experiments. The rules should be a straightforward extrapolation of Popper's ideas: If a theory makes testable and falsifiable predictions in the observable parts of the universe, we should be prepared to accept its predictions in those parts of the universe (or multiverse) that are not accessible to direct observations.

The second main reason for the hostile passions that anthropic reasoning provokes is that to some scientists it signals the "end of physics." Following Descartes, most physicists dream, above all, of a uniquely self-consistent mathematical theory that explains and determines all the microphysical constants, as well as the entire cosmic evolution. Consequently, they would like to pursue, in the words of cosmologist Edward Milne, "a single path towards the understanding of this unique entity, the universe." There is very little doubt that this was Einstein's hope too. In a lecture delivered at Oxford in 1933, Einstein said, "It is my conviction that pure mathematical construction enables us to discover the concepts and the laws connecting them which give us the key to the understanding of the phenomena of Nature." As is well known, Einstein was uncomfortable even with the probabilistic nature of quantum mechanics, though he appreciated fully its successes. In a letter he wrote on December 4, 1926, to Max Born, one of the founding fathers of quantum mechanics, Einstein expressed his opinion:

> Quantum mechanics is certainly imposing. But an inner voice tells me *that this is not yet the real thing* [emphasis added]. The theory yields much, but it hardly brings us closer to the Old One's secrets. I, in any case, am convinced that He does not play dice.

The concept of accidental variables in a potentially unobservable multiverse would have probably distressed Einstein even more.

Note, however, that Einstein's reservations about quantum mechanics stemmed more from psychology—his belief that he knew the direction in which to look—than from hard-core physics. The same may turn out to be the case with the objections to anthropic reasoning. In spite of the experience in the past few centuries, there are no assurances that physical reality will indeed oblige and render itself in its entirety to first-principles explanations. The quest for such descriptions may prove as futile as Kepler's quest for a beautiful geometrical model for the solar system. What we have traditionally called fundamental constants and maybe even laws of nature could turn out to be mere accidental variables and parochial bylaws in our universe. The anthropic principle could perhaps eventually play a similar role to that assigned by philosopher Bertrand Russell to philosophy: "The point of philosophy is to start with something so simple as to seem not worth stating, and to end with something so paradoxical that no one will believe it."

The anthropic thinking on the nature of the cosmological constant demonstrates the profound impact that Einstein's seemingly innocent attempt at a static universe continues to have on cutting-edge physics. How then do we appraise Einstein's "biggest blunder" today?

The Second Annus Mirabilis

The year 1905 is often called Einstein's annus mirabilis ("miracle year"), since during that year he published his trailblazing papers on how light knocks electrons off metals (the "photoelectric effect," which spawned quantum mechanics and won him the Nobel Prize), on random drifting of particles suspended in a fluid ("Brownian motion"), and on the theory of "special relativity." While 1905 was indeed a year of wonders for Einstein, he actually had a second annus mirabilis (fifteen months, to be exact) from November 1915 to February 1917. During this period, he published no fewer than fifteen treatises, including the brilliant pinnacle of his work—general relativity—and two significant contributions to quantum mechanics. Modern cosmology, and with it the cosmological constant, were born.

I hope that the evidence presented in chapter 10 has convinced the reader that most probably Einstein never used the expression "biggest blunder." Moreover, the introduction of the cosmological constant was not a blunder at all, since the principles of general relativity gave the green light for such a term. Thinking that the constant would ensure a static universe was definitely a regrettable mistake but not one that would qualify as a "blunder" of the magnitude considered in this book. Einstein's true blunder was to *remove* the cosmological constant! Mind again that removing the term from the equations is tantamount to arbitrarily assigning the value *zero* to lambda. In doing so, Einstein restricted the generality of his theory—a high price to pay for the conciseness of the equations, even before the recent discovery of the cosmic acceleration.

Simplicity is a virtue when it applies to the fundamental principles, not to the form of the equations. In the case of the cosmological constant, Einstein mistakenly sacrificed generality on the altar of superficial elegance. A simple analogy can help to clarify this concept. When Kepler discovered that the planetary orbits were elliptical rather than circular, the great Galileo Galilei refused to believe it. Galileo was still prisoner to the aesthetic ideals of antiquity, which assumed that the orbits had to be perfectly symmetrical. Physics has shown however that this was an unjustified prejudice. The symmetry involved actually runs much deeper than the mere symmetry of shapes. Newton's law of universal gravitation states that the elliptical orbits (which are a natural consequence of this law) can have any orientation in space. In other words, the law doesn't change whether we measure directions with respect to the north, south, or the nearest star—it is symmetric under rotations. When Einstein dubbed the cosmological constant "ugly," he was equally biased and shortsighted. He should have instead stuck with his initial instinct that a day would come "for us to be able to decide empirically the question of whether or not Λ vanishes," as he wrote to de Sitter. That day came in 1998.

Mistakes of Genius

More than 20 percent of Einstein's original papers contain mistakes of some sort. In several cases, even though he made mistakes along the way, the final result is still correct. This is often the hallmark of truly great theorists: They are guided by intuition more than by formalism. In a letter he wrote on February 3, 1915, to the Dutch physicist Hendrik Lorentz, Einstein provided his own perspective on mistakes in scientific theories:

A theorist goes astray in two ways:

1. The devil leads him by the nose with a false hypothesis. (For this he deserves our pity.)
2. His arguments are erroneous and sloppy. (For this he deserves a beating.)

Even though Einstein himself certainly committed errors of both types, his unparalleled physical insight showed him, in many cases, the path to the right answers. Unfortunately, we mere mortals can neither imitate nor acquire this talent.

In 1949 Einstein's collaborator Leopold Infeld described Einstein's pioneering paper on cosmology this way:

Although it is difficult to exaggerate the importance of this paper . . . Einstein's original ideas, as viewed from the perspective of our present day, are antiquated if not even wrong . . . Indeed, it is one more instance showing how a wrong solution of a fundamental problem may be incomparably more important than a correct solution of a trivial, uninteresting problem.

Infeld's essay appeared in a volume in Einstein's honor entitled *Albert Einstein: Philosopher-Scientist.* No fewer than six Nobel laureates in science contributed to this book. In his contribution,

Georges Lemaître described what he regarded as strong reasons for keeping the cosmological constant in the equations: "The history of science provides many instances of discoveries which have been made for reasons which are no longer considered satisfactory. It may be that the discovery of the cosmological constant [by Einstein] is such a case." How right he was.

Einstein himself remained unconvinced. In his "Remarks Concerning the Essays Brought Together in This Co-operative Volume," he repeated his earlier arguments:

> The introduction of such a constant implies a considerable renunciation of the logical simplicity of [the] theory, a renunciation which appeared to me unavoidable only so long as one had no reason to doubt the essentially static nature of space.

He continued to say that after Hubble's discovery of the cosmic expansion and Friedmann's demonstration that the expansion could be accommodated in the context of the original equations, he found the introduction of lambda "at present [in 1949] unjustified." Note, by the way, that even though Einstein wrote these comments not long after his correspondence with Gamow, there is still no allusion to "biggest blunder."

On one hand, you could argue that Einstein was right in refusing to add to his equations a term that was not absolutely required by the observations. On the other, Einstein had already missed one opportunity to predict the cosmic expansion by relying on the lack of evidence for stellar motions. By denouncing the cosmological constant, he missed a second opportunity, this time to predict the accelerating universe! With any ordinary scientist, two such oversights would surely have been regarded as lack of intuition—something we can hardly conclude about Einstein. Einstein's failures remind us that human logic is not blunder proof, even when exercised by a monumental genius.

Einstein kept thinking about a unified theory and the nature of physical reality until the end of his life. Already in 1940, he foresaw the difficulties with which current string theorists struggle: "The two systems [general relativity and quantum theory] do not directly contradict each other; but they seem little adapted to fusion into one unified theory." Then, just one month before his death, in 1955, at the age of seventy-six, he added some self-doubts: "It appears dubious whether a [classical] field theory can account for the atomistic structure of matter and radiation as well as of quantum phenomena." Einstein did, however, find some comfort in the words of the eighteenth-century dramatist Gotthold Ephraim Lessing: "The aspiration to truth is more precious than its assured possession." Blunders and all, perhaps no one in recent memory has aspired to truth more than Albert Einstein.

CODA

===

I would earnestly warn you against trying to find out the
reason for and explanation of everything . . . To try and
find out the reason for everything is very dangerous and
leads to nothing but disappointment and dissatisfaction,
unsettling your mind and in the end making you miserable.

—QUEEN VICTORIA

No scientific theory has an absolute and permanent value. As experimental and observational methods and tools improve, theories can
be refuted, or they may metamorphose into new forms that incor-
porate some of the earlier ideas. Einstein himself stressed this evolutionary nature of theories in physics: "The most beautiful fate of
a physical theory is to point the way to the establishment of a more
inclusive theory, in which it lives as a limiting case." Darwin's theory for the evolution of life by means of natural selection was only
strengthened through the application of modern genetics. Newton's
theory of gravity continues to live as a limiting case within the framework of general relativity. The road to a "new and improved" theory
is far from smooth, and progress is definitely not a headlong rush
to the truth. If luminaries such as Darwin, Kelvin, Pauling, Hoyle,
and Einstein can commit serious blunders, imagine the scorecards
of lesser scientists. When James Joyce wrote in *Ulysses,* "A man of
genius makes no mistakes. His errors are volitional and are the portals of discovery," he meant the first part of his comment to be provocative. As we have seen in this book, however, the blunders of
genius are often indeed the portals of discovery.

In director Rob Reiner's 1987 fairy-tale film *The Princess Bride,* one of the characters engages in a battle of wits against the protagonist. At one point, he exclaims, "You fell victim to one of the classic blunders! The most famous of which is 'never get involved in a land war in Asia.'" I think we can all agree that recent history has shown this statement to be good advice. The famous mathematician and philosopher Bertrand Russell suggested another tip to those who want to make sure they avoid fanaticism: "Do not feel absolutely certain of anything." The examples in this book demonstrate that this "commandment" can also be taken as a useful hint for how to dodge major blunders—but I am not absolutely certain about this . . . While doubt often comes across as a sign of weakness, it is also an effective defense mechanism, and it's an essential operating principle for science.

Kelvin, Hoyle, and Einstein revealed yet another fascinating side of human nature. Just as people (scientists included) are sometimes reluctant to admit their mistakes, they also on occasion stubbornly oppose new ideas. Max Planck, one of the forefathers of quantum mechanics, once remarked cynically, "New scientific truth does not triumph by convincing its opponents and making them see the light, but rather because its opponents eventually die, and a new generation grows up that is familiar with it." This may be sad but true.

Psychologists Amos Tversky and Daniel Kahneman established a cognitive basis for common human errors using the concept of *heuristics*: simple rules of thumb that guide decision making. One of their findings was that people tend to rely more on their intuitive understanding—which is based largely on their personal experience—than on actual data. Naturally, scientists of the caliber of Darwin, Pauling, or Einstein believed that their intuition would guide them to the correct answer even when the right way forward was elusive or when the scientific landscape was changing at a bone-breaking pace. As I have noted above, Bertrand Russell understood the dangers of overconfidence and certainty, and he thought that he had found a solution when he advocated a habit of hinging beliefs "upon observations and inferences as impersonal, and as much divested of local and temperamental bias, as is possible for human beings." Unfortunately, it is not easy to follow this advice. Modern neuroscience has shown unambiguously that the orbitofrontal cortex (a region in the frontal lobes in the brain) integrates emotions into the stream of rational thought. Humans are not purely rational beings capable of completely turning off their passions.

Despite their blunders, and perhaps even *because* of them, the five individuals I have followed and sketched in this book have produced not just innovations within their respective sciences but also truly great intellectual creations. Unlike many scientific works that target only professionals from within the same discipline as their audience, the oeuvres of these masters have crossed the boundaries between science and general culture. The impact of their ideas has been felt far beyond their immediate significance for biology, geology, physics, or chemistry. In this sense, the work of Darwin, Kelvin, Pauling, Hoyle, and Einstein comes closer in spirit to achievements in literature, art, and music—both cut a broad swath across erudition.

There is no better way to end a book on blunders than with an important reminder—a plea for humility, if you like—that nobody can express more eloquently than Darwin:

> We must, however, acknowledge, as it seems to me, that man with all his noble qualities, with sympathy which feels for the most debased, with benevolence which extends not only to other men but to the humblest living creature, with his god-like intellect which has penetrated into the movements and constitution of the solar system—with all these exalted powers—Man still bears in his bodily frame the indelible stamp of his lowly origin.

NOTES

Chapter 1: Mistakes and Blunders

PAGE

5 *world championship match:* A detailed description can be found in Evans and Smith (1973). A brief summary is online at www.mark-weeks.com/chess/72fs$$.htm.

6 *Ray Krone of Phoenix:* A description of this sad case can be found online, eg, at www.innocence project.org/Content/Ray_Krone.php.

7 *The British historian A. J. P. Taylor:* Alan John Percival Taylor (1906–90). Taylor 1963.

9 *President Woodrow Wilson:* Wilson 1913.

Chapter 2: The Origin

PAGE

12 *On November 29, 1975:* The record is apparently held by the vulture known as Rüppell's Griffon (*Gyps rueppellii*); see www.straightdope.com/columns/read/1976/how-high-can-birds-and-bees-fly.

13 *record-setting explorer Jacques Piccard:* See, eg, "Jacques Piccard," *Encyclopedia of World Biography,* 2004. Online at www.encyclopedia.com/doc/1G2-3404707243.html.

13 *A recent catalogue:* Chapman 2009.

13 *The most recent study predicts:* Mora et al. 2011.

13 *one tablespoon of dirt:* Gans et al. 2005.

14 *The ocellaris clown fish:* Scientifically, *Amphiprion ocellaris.*

14 *According to Greek philosopher Aristotle:* Aristotle fourth century BCE.

14 *A similar description appears:* Pliny the Elder first century CE. Can be downloaded from www.perseus.tufts.edu.

15 *The famous Roman orator:* Cicero 45 BCE.

15 *precisely the line of reasoning adopted by Paley:* Paley 1802. William Paley (1743–1805) published an influential book entitled *Natural Theology,* in which he contrasted a natural rock with a watch. Ironically, through radiometric dating (see chapter 5), rocks can determine the age of the Earth—a

much longer time interval than that ever measured by any watchmaker's clock.

17   *The first edition of Darwin's book:* There are, of course, numerous print-ings of *The Origin.* Two that I found particularly attractive are *The Annotated Origin,* annotated by James T. Costa (Darwin 2009), and one reprinted in facsimile with an introduction by Ernst Mayr (Darwin 1964).

18   *"In the distant future I see":* Darwin 2009 [1859], p. 488. Darwin himself followed up on his own prediction in *The Descent of Man,* published in 1871, and in *The Expression of the Emotions in Man and Animals,* pub-lished the following year. Current developments in evolutionary psychol-ogy can be seen as descendants of these pioneering efforts.

18   *"We thus learn that man":* Darwin 1981 [1871]. A dozen years following *The Origin,* Darwin became confident enough to broaden his theory of evolution to include humans—an issue he tried to skirt in *The Origin.* There is very little doubt that the outcry against Darwinism would have been much less pronounced had evolution not applied to humans. Dar-win's ideas in *The Descent* have inspired tireless efforts by many members of the Louis Leakey family to search for and find fossil hominids in Africa.

18   *Darwin's theory consists of four main pillars:* There are many excellent books on evolution and natural selection, at various levels. Here are just a few that I found very helpful: Ridley 2004a is a first-class textbook. Ridley 2004b is a wonderful anthology of high-level articles, as are Hodge and Radick 2009 (on Darwin), and Ruse and Richards 2009 (on *The Origin*). A thought-provoking philosophical approach is Dennett 1995. An excellent review of the history of evolutionary theory is Depew and Weber 1995. Wilson 1992 is a sweeping review of biodiversity. Dawkins 1986, 2009, Carroll 2009, and Coyne 2009 are superb popular books. Pallen 2009 is a brief and extremely accessible introduction. A few very useful websites on evolution are www.evolution.berkeley.edu; www.pbs.org/evolution; and www.nationalacademies.org.evolution.

20   *Darwin's ideas on evolution had an older:* A landmark work on the history and origins of the theory of evolution is Gould 2002. Another high-level historical review is Bowler 2009.

20   *distinguish between* microevolution: Resistances to antibiotics and to pes-ticides, which develop within a few years, are examples of microevolution. The origin of mammals from reptiles is an example of macroevolution. An excellent summary of macroevolution is Carroll, Grenier, and Weatherbee 2001.

20   *"Nothing in Biology Makes Sense":* Dobzhansky 1973.

21   *modern theory of* uniformitarianism: Charles Lyell (1797–1875) largely expanded on the concept that geological changes are the result of the con-tinuous accumulation of tiny transformations over immeasurably long periods of time, in his influential book *Principles of Geology.* Lyell 1830–33.

21 *Some "living fossils" such as the lamprey:* Classified as *Priscomyzon rinien-sis*. Gess et al. 2006.

21 *the concept of a* common ancestor: This pillar of Darwinian evolution has been confirmed by many spectacular findings. For instance, the discoveries of fossils of feathered dinosaurs, such as the *Microraptor gui* and the *Mei long*, are consistent with the idea that birds evolved from reptiles.

22 *Darwin's solution to the diversity problem:* Good descriptions of speciation are Schilthuizen 2001, and Coyne and Orr 2004.

22 *the "tree of life":* An interesting discussion on the tree of life can be found in Dennett 1995.

22 *the case of the Italian sparrow:* Elgvin et al. 2011.

23 *author Vladimir Nabokov:* The study that confirmed Nabokov's speculation is Vila et al. 2011.

23 *phylogenetic tree for all the families:* In an impressive study, Meredith et al. 2011 used twenty-six genes to construct the phylogeny of mammalian families and to estimate divergence times.

23 The Accelerating Universe: Livio 2000.

24 *I mean reductionism:* This often-abused term is sometimes used incorrectly to imply that one can ignore complexities and completely reduce one discipline into another. No one should attempt to understand Lord Byron's "Don Juan" from the perspective of the laws of physics. A good discussion of reductionism in the sense that I am using it here can be found in Weinberg 1992.

26 *the British geneticist J. B. S. Haldane is cited:* Eg, Hutchinson 1959.

26 *a fresh water ameboid:* Given that the genome determination was done with earlier methods, it may be somewhat uncertain; McGrath and Katz 2004.

27 *The basic idea underlying natural selection:* A comprehensive book is Bell 2008. See also Endler 1986.

27 *In a letter to Darwin:* Darwin and Seward 1903.

27 *Here is how Darwin himself:* Darwin 1964 [1859], p. 61.

28 *by the term of Natural Selection:* A highly accessible description of natural selection can be found in Mayr 2001. A textbook on selection is Bell 2008. Endler 1986 presents much of the evidence for natural selection.

29 *Wallace wrote to Darwin:* Marchant 1916, p. 171.

29 *philosophical radicals such as the political economist Thomas Malthus:* Malthus argued in his *An Essay on the Principle of Population* (published in 1798) that humans produce too many offspring and that, consequently, if unchecked, famine and "premature death must in some shape or other visit the human race." Malthus's ideas influenced not just Darwin and Wallace but also economic and political philosophy.

31 *Darwin himself wrote to the geologist:* The geologist Frederick Wolaston Hutton (1836–1905) reviewed *The Origin* for the *Geologist*.

31 *no fewer than a half million patients:* Bowersox 1999.

32 *the evolution of the peppered moth:* The British geneticist Bernard Kettlewell (1907–79) conducted much research on the peppered moth and industrial melanism. His findings have been questioned by some (eg, Wells 2000; Hooper 2003), and supported by others (eg, Majerus 1998). A popular-science summary of the debate is de Roode 2007.

33 *the famous philosopher of science:* Popper 1976, p. 151.

33 *did recognize his error:* Popper 1978; also Miller 1985.

33 *termed by modern evolutionary biologists* genetic drift: There exists vast literature on genetic drift. An online lecture by Stephen Stearns can be found at www.cosmolearning.com/video-lectures/neutral-evolution -genetic-drift-66 87. Other easy-to-access online resources include Kliman et al. 2008, and www.ucl.ac.uk/~ucbhdjm/courses/b242/InbrDrift/ InbrDrift.html. A comprehensive textbook on population genetics is Hartl and Clark 2006.

34 *Ellis-van Creveld syndrome:* This is a manifestation in the Amish community of what is known as the "founder effect." When a population is reduced to a very small size because of some environmental changes or because of migration, the genes of the "founders" of the resulting population are represented disproportionately.

35 *suffragist and botanist Lydia Becker:* Lydia Ernestine Becker (1827–90) published the *Women's Suffrage Journal* between 1870 and 1890. The citation on Darwin is from her address, as president of the Manchester Ladies' Literary Society, on January 30, 1867. It was published in Becker 1869. It is also described in Blackburn 1902, part 2.

Chapter 3: Yea, All Which It Inherit, Shall Dissolve

PAGE

37 *confessed candidly in* The Origin: Darwin 2009, p. 13.

38 *In this "paint-pot theory":* This phrase was first used by Hardin 1959, p. 107.

38 *"After twelve generations":* Darwin 2009, p. 160.

38 *Jenkin was a multitalented individual:* Brownlie and Lloyd Prichard 1963.

39 *Fleeming Jenkin published his criticism:* Jenkin 1867. The article is reproduced in Hull 1973, p. 303, and can also be found online at www.victorian web.org/science/science_texts/jenkins.html.

39 *Jenkin's approach was quantitative:* Excellent discussions of Jenkin's arguments can be found in Bulmer 2004, Vorzimmer 1963, and Hull 1973.

40 *only Arthur Sladen Davis:* Davis 1871.

42 *The modern theory of genetics:* An engaging description of Mendel and his work can be found in Mawer 2006.

42 *A simple example will help to clarify:* The description here is largely a simplified version of the explanation in Ridley 2004a, pp. 35–39.

44 *it needed Mendelian heredity:* First explained in detail in Fisher 1930.

44 *In his autobiography, he acknowledged:* Darwin 1958 [1892], p. 18. A more detailed analysis of Darwin's numerical attempts can be found in Parshall 1982.

47 *"I was blind and thought that single":* Letter to Wallace on February 2, 1869, in Marchant 1916, vol. 1. Also in Darwin 1887, vol. 2, p. 288.

47 *"If in any country or district":* Darwin 1909 [1842], p. 3.

47 *In reality, Darwin even relied:* Hodge 1987.

47 *"When a character which has been lost":* Darwin 2009 [1859], p. 160.

47 *This notion of some latent "tendency":* Darwin returned to this notion of a latent tendency in a letter he wrote to Wallace on September 23, 1868 (Darwin and Seward 1903, vol. 2, p. 84). Darwin wrote: "I think impossible to see how, for instance, a few red feathers appearing on the head of a male bird, and which are at first transmitted to both sexes, would come to be transmitted to males alone. It is not enough that females should be produced from the males with red feathers, which should be destitute of red feathers; but these females must have a latent tendency to produce such feathers, otherwise they would cause deterioration in the red head-feathers of their male offspring."

48 *In a letter to Wallace on January 22:* Darwin was working on the fifth edition of *The Origin;* in F. Darwin 1887, vol. 3, p. 107. See also Bulmer 2004, Morris 1994.

48 *"I saw, also, that the preservation":* Peckham 1959, p. 178.

49 *of which* two *survive to reproduce:* Peckham 1959, p. 178.

49 *"Approaching the subject [of evolution]":* The precise date of this letter is not known, but because it was sent from Moor Park, it had to be before November 12, 1857. In Darwin and Seward 1903, vol. 1, p. 102.

50 *he wrote in his book* The Variation: Darwin 1868, vol. 2, p. 374.

51 *First, in a letter written on January 22:* Marchant 1916, vol. 1, p. 166.

51 *Wallace replied on February 4:* Marchant 1916, vol. 1, p. 168.

51 *Darwin was quick to correct Wallace:* Letter dated "Tuesday, February, 1866." Marchant 1916, vol. 1, p. 159.

52 *Darwin points out the obvious fact:* The exchange between Darwin and Wallace and its significance is also discussed beautifully in Dawkins 2009.

52 *Gregor Mendel read the seminal paper:* In addition to Mawer 2006, Orel 1996 gives a detailed account of Mendel's life and work. See also Brannigan 1981.

52 *no fewer than three books:* Kitcher 1982, p. 9; Rose 1998, p. 33; Henig 2000, p. 143–44.

52 *and a fourth book:* Dover 2000, p. 11.

52 *Andrew Sclater of the Darwin Correspondence:* Sclater 2003. See also Keynes 2002.

54 *In contrast to Darwin's total lack:* An excellent description of the influences (or lack thereof) between Darwin and Mendel is de Beer 1964.

55 *"If the change in the conditions":* Mendel 1866 [1865], p. 36 (cited in de Beer 1964).

55 *"No case is on record":* Darwin 1964 [1859], p. 7; or Darwin 2009, p. 8.

56 *"If it be accepted that":* Mendel 1866, p. 39 (cited in de Beer 1964).

57 *Although the Vatican itself:* For a discussion of the early Vatican responses to evolution, see Harrison 2001.

57 *the illusion of confidence:* The effect was demonstrated by Kruger and Dunning 1999. A popular description can be found in Chabris and Simons 2010.

Chapter 4: How Old Is the Earth?

PAGE

60 *the Hindu sages of antiquity:* The ancient Hindus believed that a single cycle of destruction and renewal lasted 4.32 million years (eg, Holmes 1947, p. 99–108).

61 *In one of the earliest estimates:* Theophilus of Antioch (ca. 115–180 CE) was converted to Christianity as an adult. Only one of his writings has survived, in an eleventh-century manuscript, quoted in Haber 1959, p. 17, and in Dalrymple 1991, p. 19.

61 *among these biblical scholars were John Lightfoot:* Ussher (1581–1656) calculated that Creation occurred in the year 710 of the Julian period; Brice 1982.

61 *added as a marginal note to the English:* The note was removed at the beginning of the twentieth century. Kirkaldy 1971, p. 5.

62 *"The knowledge of . . . nearly everything":* Spinoza 1925, vol. 3, p. 98.

62 *"It would be a mark of great":* Philo, first century, book I.

62 *He pointed out in 1754:* Kant 1754. An English translation appears in Reinhardt and Oldroyd 1982.

63 *Referring to a sarcastic passage:* The reference is to Fontenelle's *A Plurality of Worlds.*

63 *Benoît de Maillet carried out:* An English translation of de Maillet's 1748 book is Carozzi 1969.

63 *"Why the bones of great":* MacCurdy 1939, p. 342.

64 *De Maillet humbly dedicated:* de Maillet 1748; Cyrano de Bergerac was the author of the imaginative two-volume *Travels to the Moon and the Sun.*

64 *"a globe of red hot iron":* Newton 1687; see the English translation by Motte 1848, p. 486.

65 *Buffon assumed that the Earth started:* The twentieth volume in Buffon's *Natural History: General and Particular* was entitled *Epochs of Nature.* In

it, he divided the history of the Earth into seven epochs, and he attempted to estimate the length in time of each. A good description can be found in Haber 1959, p. 118.

66 *"no vestige of a beginning":* Hutton 1788.

66 *"how fatal the suspicion":* Richard Kirwan was president of the Royal Irish Academy. He wrote a series of articles and a book in support of the biblical description and against Hutton. The quote here is from Kirwan 1797.

66 *with the publication of Charles Lyell's:* Lyell 1830–33.

67 *published a eulogy of Lord Kelvin:* There are quite a few detailed biographies of Lord Kelvin. The ones I found most illuminating are Gray 1908, Thompson 1910 (reprinted in 1976), Smith and Wise 1989, Lindley 2004, and Sharlin and Sharlin 1979. Wilson 1987 describes side by side the physics of Kelvin with that of Victorian physicist Sir George Gabriel Stokes (Stokes lived 1819–1903, Kelvin 1824–1907). Burchfield 1990 concentrates on Kelvin's work on the problem of the age of the Earth.

68 *"after making himself Second Wrangler":* The "Senior Wrangler" was the undergraduate student who received the highest grades in the mathematical honors examinations at Cambridge, known as the "Tripos." William Thomson was expected by most to be Senior Wrangler. In fact, his tutor, Dr. Cookson, noted that it would be "a great surprise to the University if he were not." Thomson himself was somewhat less convinced. When the competition began, another student, Stephen Parkinson, who was apparently more efficient in answering very rapidly and economically, emerged as a candidate to win. At the end, the more talented but less speedy Kelvin indeed came in second. Thomson did beat Parkinson, however, for the Smith's Prize, awarded for the best performance in a series of examinations and thought to require more profound analytical understanding.

69 *"I may say that the one thing":* Kelvin made this comment in his Baltimore Lectures on molecular dynamics and the wave theory of light, given at the Johns Hopkins University in 1884.

69 *"On the Secular Cooling":* Kelvin 1864.

70 *"On the Age of the Sun's":* Kelvin 1862.

70 *"For eighteen years it has pressed":* Kelvin 1864. The paper was read on April 28, 1862.

70 *Kelvin's first papers on the topics of heat:* Kelvin made numerous contributions to thermodynamics. In 1844 he had a paper on the "age" of temperature distributions. Basically, he showed that a temperature distribution measured at the present can be the result only of a heat distribution that existed some finite time into the past. In 1848 he devised the absolute temperature scale that bears his name. In an 1851 paper entitled "On the Dynamical Theory of Heat," he formulated one version of what is known today as the second law of thermodynamics.

71 *"To suppose, as Lyell":* Kelvin 1864.

71 *the French physicist Joseph Fourier:* A good description of the development of the theory of thermal conductivity can be found in Narasimhan 2010.

73 *Kelvin believed that he could state:* Kelvin admitted that "we are very ignorant as to the effects of high temperatures in altering the conductivities and specific heats of rocks, and as to their latent heat of fusions." These types of uncertainties ended up playing an important role in his blunder.

74 *Kelvin was able to obtain a rough estimate:* This timescale is known today as the Kelvin-Helmholtz timescale.

74 *"It seems, therefore, on the whole":* Kelvin 1862. Shaviv 2009 presents a very detailed but quite accessible exposition of the theory of stellar structure and evolution.

75 *Kelvin later described the conversation:* Thomson 1899. Chamberlin 1899 presents a commentary on Kelvin's address in 1899.

76 *when Kelvin delivered an address:* On February 27, 1868; Kelvin 1891–94, vol. 2, p. 10.

77 *"The earth, if we bore into it":* Kelvin 1891–94, vol. 2, p. 10.

78 *Since tides caused by the Moon's:* The Earth's angular velocity in its spin around its axis is higher than the angular velocity of the Moon in its orbit. Consequently, the tidal forces tend to spin the Earth down and to increase the Earth-Moon distance.

78 *"It is impossible, with the imperfect data":* Kelvin 1868.

78 *George was a physicist:* While in Trinity College in Cambridge, George Darwin (1845–1912) was Second Wrangler and Second Smith's Prizeman.

78 *even a solidified Earth was not:* Darwin repeated the result of his 1878 letter considering the rigidity of the Earth in his presidential address to the British Association in 1886 (G. H. Darwin 1886), concluding that he saw no right to be "so confident of the internal structure of the earth as to be able to allege that the earth would not through its whole mass adjust itself almost completely to the equilibrium figure."

79 *Charles Darwin was delighted:* G. H. Darwin in Stratton and Jackson 1907–16, vol. 3, p. 5.

79 *"We, the geologists [emphasis added], are at fault":* Kelvin 1891–94, vol. 2, p. 304.

80 *"always felt that this hypothesis":* Kelvin gave his presidential address entitled "On the Origin of Life" at Edinburgh in August 1871. Kelvin 1891–94, vol. 2, p. 132.

80 *"profoundly convinced that the argument":* Kelvin 1891–94, vol. 2, p. 132.

80 *started to pay serious attention to Kelvin's:* Burchfield 1990 (especially chapters 3 and 4) provides a comprehensive discussion of Kelvin's influence and impact.

81 *He is perhaps best known for his legendary:* The event took place during the meeting of the British Association for the Advancement of Science,

held its thirtieth annual conference from June 27 to July 4, 1860. ...main event on June 30 was a rather long lecture by historian of science ...n William Draper. Estimates of the attendance (in the *Evening Star*) ...at it somewhere between four hundred and seven hundred (issue of July 2). Bishop Wilberforce's comments following the lecture apparently lasted for about a half hour. He concluded that "Mr. Darwin's conclusions were an hypothesis, raised most unphilosophically to the dignity of a causal theory" (as reported on July 7 in the *Athenaeum*). The most thorough analysis of the details of the event is in Jensen 1988. See also Lucas 1979.

81   *The story was told in colorful:* Sidgwick 1898.

82   *Even though there are many versions:* For example, the *Press* reported on July 7: "[The Bishop] asked the professor [Huxley] whether he would prefer a monkey for his grandfather or his grandmother." Huxley himself had written to his friend Dr. Frederick Dryster on September 9, 1860: "except indeed the question raised as to my personal predilections in the matter of ancestry . . . If then, said I, the question is put to me would I rather have a miserable ape for a grandfather or a man highly endowed by nature and possessed of great means and influence and yet who employs those faculties for mere purpose of introducing ridicule into a grave scientific discussion—I unhesitatingly affirm my preference for the ape." The letter is in *The Huxley Papers*, 15, 117 (London: Imperial College); it is cited in Foskett 1953.

82   *Historian of science James Moore:* Moore 1979, p. 60.

82   *"I do not suppose that":* Huxley 1909 [1869], p. 335–36.

83   *"We find that we may":* Tait 1869.

83   *to name the ten greatest:* The list that appeared in the December 1999 issue of *Physics World* included, in this order: Albert Einstein, Isaac Newton, James Clerk Maxwell, Niels Bohr, Werner Heisenberg, Galileo Galilei, Richard Feynman, Paul Dirac, Erwin Schrödinger, and Ernest Rutherford. Slightly different lists appeared in other polls. (In particular, on several lists, Newton was ranked first, with Einstein second.)

83   *We know today that the age:* See, eg. Dalrymple 2001.

Chapter 5: Certainty Generally Is Illusion

PAGE

84   *"Mathematics may be compared":* Huxley 1909 [1869].

85   *the engineer John Perry:* John Perry (1850–1920) was born in Ireland. After being a professor of mechanical engineering in both the United Kingdom and Japan, he was appointed professor of engineering and mathematics at Finsbury Technical College in London. In 1896 he advanced to a professorship at the Royal College of Science. Throughout his career, Perry introduced novel teaching methods in mathematics and worked on prob-

lems of applied electricity. See, eg, Nudds, McMillan, Weaire, and McK-
enna Lawlor 1988; Armstrong 1920.

85 *to argue that evolution by natural selection:* Salisbury argued that one hundred million years was not sufficient for natural selection to transform
jellyfish into humans, and also repeated Kelvin's objection based on the
argument of design. Salisbury 1894. Described also in Shipley 2001.

85 *Perry wrote to a physicist friend:* Perry wrote to Oliver Lodge on October 31, 1894. He added that if natural selection were to be dismissed, the
only alternative was to appeal to some providence, and he regarded that as
destructive for scientific reasoning. Shipley 2001.

85 *he diligently sent copies:* To physicists Joseph Larmor and George FitzGerald, as well as to Osborne Reynolds and Peter Guthrie Tait. He also wrote
to Kelvin on October 17, 1894, and again on October 22 and October
23 (Cambridge University Library, Papers of Lord Kelvin Add. MS. 7342
(P56, P57, P58).

86 *"I sat beside him":* The dinner took place on October 28, 1894. Perry
wrote to Oliver Lodge on October 29 (University College London, Lodge
Papers Add. MS. 89).

86 *The scientific journal* Nature *eventually published:* Perry 1895a.

87 *In an offensively dismissive letter:* An extract from Tait's letter to Perry is
included in Perry 1895a.

88 *"You say I am right":* Letter of Perry to Tait on November 26, 1894. This
letter is also included in Perry 1895a. Perry also noted, "I found that so
many of my friends agreed with me."

88 *"I should like to have your":* Letter of Tait to Perry on November 27, 1894
(Cambridge University Library, Papers of Lord Kelvin, Add. MS. 7342,
P59d). Included in Perry 1895a.

88 *"It is for Lord Kelvin to prove":* Letter of Perry to Tait on November 29,
1894. Included in Perry 1895a. Perry emphasized two arguments in his letter: (1) that there was a certain amount of fluidity inside the Earth, so that
heat could be transported by convection, and (2) that according to results
by Robert Weber, the conductivity of rocks increased with increasing temperature. Later, the latter turned out to be wrong.

89 *"I feel that we cannot assume":* Letter from Kelvin to Perry on December
13, 1894. Included in Perry 1895a, p. 227. Kelvin was interested in particular in checking Weber's results on the conductivity.

89 *"refusing sunlight for more":* Kelvin to Perry on December 13, 1894.
Included in Perry 1895a.

90 *Perry's challenge caused:* On his part, Perry asked for the assistance of
mathematician Oliver Heaviside, and he published a more sophisticated
mathematical analysis of the problem; Perry 1895b.

90 *The overjoyed Kelvin published:* Thomson (Lord Kelvin) 1895, Thomson (Lord Kelvin) and Murray 1895. Kelvin based his conclusions also on

measurements of the melting point of diabase, a basalt, by geologist Carl Bakus.

91 *Perry concluded his last:* Perry 1895c. The entire debate is described in detail in Shipley 2001 and Burchfield 1990.

92 *The phenomenon became known as* radioactivity: The discovery of radioactivity is in Becquerel 1896; the discovery that heat is produced is in Curie and Laborde 1903.

92 *It took the amateur astronomer:* Wilson 1903. His letter to the editor was only fifteen lines in length.

92 *did not escape George Darwin:* Darwin 1903. He speculated that Kelvin's estimated age could be increased by a factor of ten or twenty.

92 *The Irish physicist and geologist:* Joly 1903.

93 *New Zealand–born physicist Ernest Rutherford:* A good biography of Rutherford and a description of his work is Eve 1939.

93 *Kelvin showed great interest:* He discussed it in a letter to the English physicist Lord Rayleigh on August 24, 1903, and also with Rutherford himself and with Pierre and Marie Curie on their visit to England.

93 *"I venture to suggest that somehow":* Kelvin 1904.

93 *In 1904, however, with considerable:* Physicist and Nobel laureate Sir Joseph John "J. J." Thomson (not related to Lord Kelvin), who discovered the electron, reminisced in 1936 that Kelvin acknowledged in a conversation that the discovery of radioactive heating undermined his assumptions in the calculation of the Earth's age; Thomson 1936, p. 420. Kelvin made a similar concession during the British Association meeting; Eve 1939, p. 109.

93 *In an acerbic exchange:* It started with a letter by Kelvin, published on August 9, 1906, in which he repeated his belief that the Sun's energy was only gravitational and asserted that radioactivity was no more than a hypothesis. Various rebuttal letters by Frederick Soddy, Oliver Lodge, Robert John Strutt (physicist and son of Lord Rayleigh), and Kelvin appeared for about a month. In his letter of August 15, Lodge said about Kelvin that "his brilliantly original mind has not always submitted patiently to the task of assimilating the work of others by the process of reading." The episode is described briefly in Eve 1939, pp. 140–41, Burchfield 1990, p. 165, and Lindley 2004, p. 303. A review of the controversy is in Soddy 1906.

94 *I came into the room:* Cited, eg, in Eve 1939, p. 107.

94 *Eventually,* radiometric dating: Holmes 1947 gives a nice review. Today's accepted age was first determined by geochemist Clair Patterson by using data from the Canyon Diablo meteorite (Patterson 1956). Scientists at Argonne National Laboratory have put radiometric dating to another interesting use. Using the decay of the rare isotope krypton 81, they succeeded in 2011 in tracking the ancient Nubian Aquifer that stretches across northern Africa.

94 *Rutherford was walking on campus:* Eve 1939, p. 107.

95 *Geologists Philip England, Peter Molnar:* England, Molnar, and Richter 2007; Richter 1986.

96 *The theory of* cognitive dissonance: The classical text is Festinger 1957. More recent studies have revealed complex details, both in the psychological and neuroscience arenas; eg, Cooper and Fazio 1984, vol. 17, p. 229; Lee and Schwartz 2010; Van Overwalle and Jordens 2002; Van Veen et al. 2009.

96 *The messianic stream within the Jewish:* Interesting descriptions and an analysis of the events surrounding Schneerson's death can be found in Ochs 2005 and Dein 2001.

97 *An experiment conducted in 1955:* Brehm 1956.

98 *Already in the 1950s, researchers:* Olds and Milner 1954; Olds 1956 is a popular version.

98 *Studies showed that an important part:* There have been many studies of positive affective reactions and of addictions. See, eg, Bozarth 1994; Fiorino, Coury, and Phillips 1997; Berridge 2003; and Wise 1998. A popular-science account is Nestler and Malenka 2004, and very accessible popular books on the experience of pleasure are Linden 2011 and Bloom 2010.

99 *Neuroscientist and author Robert Burton suggested specifically:* Burton 2008, pp. 99–100, and p. 218.

99 *is not associated with neural activity:* Motivated reasoning implies an emotion regulation. The studies suggest that motivated reasoning is qualitatively different from reasoning when people do not have a strong emotional stake in the results. An extensive review on motivated reasoning is Kunda 1990. The involvement of emotion in decision making is discussed, eg, in Bechara, Damasio, and Damasio 2000. A popular account is Coleman 1995. Westen et al. 2006 present the fMRI studies.

100 *"The concordance of results":* King 1893.

100 *his objection to revising:* A good discussion of the importance of Kelvin's estimate for the age of the Sun is in Stacey 2000.

100 *In August 1920:* I discuss the problem of the generation of energy in stars in chapter 8.

101 *"With respect to the lapse of time":* Darwin inserted this sentence in the sixth edition; Peckham 1959, p. 728.

Chapter 6: Interpreter of Life

PAGE

103 *The lecture hall in the Kerckhoff:* Hager 1995, p. 374, gives a nice description of the event.

103 *Watson was visiting the Swiss:* Watson was on his way back from Naples

to Copenhagen, Denmark, where he was a postdoctoral fellow, and he stopped in Geneva.

104   *had made it even into the pages:* "Chemists Solve a Great Mystery: Protein Structure Is Determined," *Life*, September 24, 1951, pp. 77–78.

104   *Pauling started to think about proteins:* There are quite a few biographies of Pauling. I found the following particularly helpful: Hager 1995; Serafini 1989; Goertzel and Goertzel 1995; and Marinacci 1995. A number of books cover various aspects of Pauling's work excellently. Among them I would like to mention Olby 1974; Lightman 2005; Judson 1996; and, of course, the fantastic website at Oregon State University: http://osulibrary .oregonstate.edu/specialcollections//coll/pauling.

104   *His first papers on the subject:* Pauling 1935; Pauling and Coryell 1936. Pauling and chemist Charles D. Coryell performed the experiment by suspending between the poles of a large magnet a tube of cow blood; Judson 1996, pp. 501–2, gives a good description.

104   *Alfred Mirsky, a leading protein expert:* Pauling did not have much expertise with protein molecules, so he convinced Mirsky, who was at the Rockefeller Institute for Medical Research, to come to Caltech for the 1935–36 year. (He also convinced the Rockefeller Institute's president to allow Mirsky to leave!)

105   *Mirsky and Pauling first proposed:* Mirsky and Pauling 1936. Some earlier work was done by Hsien Wu in 1931.

105   *is composed of chains:* Very significantly for Pauling's subsequent work, the authors noted that "this chain is folded into a uniquely defined configuration, in which it is held by hydrogen bonds." Hydrogen bonds—where the hydrogen is held jointly by two atoms, effectively creating a bridge between them—were about to become Pauling's trademark.

105   *obtained by the physicist William Astbury:* Astbury 1936.

105   *Pauling immersed himself in the work:* Pauling described his activities at the time in a dictation given in 1982. The transcription was published by Pauling's assistant, Dorothy Munro; Pauling 1996.

105   *Figure 11 shows a schematic drawing:* Pauling's original piece of paper, on which he sketched the structure and then folded it, in 1948, was never discovered.

106   *Pauling convinced Robert Corey:* Corey had considerable experience with X-ray studies of proteins already. Many years later, Pauling commented graciously that it may actually have been Corey who convinced him.

107   *"In the spring of 1948":* Pauling 1996. I should note that in an earlier account, Pauling 1955, Pauling says that he found only one of the two helices in Oxford, while the other was discovered by Herman Branson when Pauling returned to Caltech.

107   *Pauling created a* helix: Olby 1974, p. 278, gives an excellent, if somewhat technical, description of the road to the alpha-helix.

109 *"They have about five times"*: Pauling wrote to chemist and crystallographer Edward Hughes. Cited on *The Pauling Blog* website, under "An Era of Discovery in Protein Structure."

109 *Even during a discussion with the famous:* Pauling admitted in later interviews that he had been concerned that the Cavendish group might beat him to the punch in checking the models. Olby 1974, p. 281; Hager 1995, p. 330.

109 *whether Branson could find:* According to Pauling 1955, Branson may have found only one of the two helices, after Pauling explained to him all the important constraints involved. In Pauling 1996, he gives the impression that he (Pauling) had discovered both helices in Oxford and that Branson later confirmed them.

110 *"Polypeptide Chain Configurations":* Bragg, Kendrew, and Perutz 1950.

110 *The idea behind X-ray crystallography:* Good descriptions of the technique itself and its applications can be found, eg, in McPherson 2003. An outline containing less physics is Blow 2002.

111 *"Proteins are built of long chains":* Bragg, Kendrew, and Perutz 1950.

111 *Bragg hammered nails:* Perutz 1987.

111 *Pauling was always extremely competitive:* Alex Rich, Jack Dunitz, and Horace Freeland Judson all confirmed this fact in conversations with the author.

112 *he and Corey sent a short note:* Pauling and Corey 1950.

112 *that contained a detailed explanation:* Pauling, Corey, and Branson 1951. Somewhat sadly, Branson wrote a letter in 1984 to Pauling's biographers Ted and Ben Goertzel, alleging that it was he, and not Pauling, who had "found two spiral structures which fit all the data." In 1995 he added that Corey had nothing to do with the discovery (Goertzel and Goertzel 1995, pp. 95–98). These allegations are inconsistent with the recollections of a number of other scientists, who remembered Pauling's models from Oxford, and also inconsistent with the fact that Branson had agreed to be third author on the paper. Branson himself did note that Pauling was "one of the impressive scientific intellects of our age who deserves the Nobel Prizes."

112 *that he thought that the word "spiral":* Dunitz, in a conversation with the author on November 23, 2010.

113 *"I was thunderstruck by Pauling":* Perutz 1987.

114 *He was one of the first scientists:* A concise summary of Pauling's achievements is Dunitz 1991.

114 *"To understand all these great biological":* Pauling 1948a.

115 *"I believe that as the methods":* Pauling 1939, p. 265.

115 *Even the space-filling colored models:* Francoeur 2001, p. 95. See also Nye 2001, p. 117.

115 *"The Gregorian monk Mendel":* Pauling 1948b.

116 *"The detailed mechanism by means":* Pauling 1948b.

117 *Levene managed to distinguish:* eg, Levene and Bass 1931. Olby 1974, p. 73–96 gives a good description of the early work.

117 *"The nucleic acids of the nucleus":* Wilson 1925.

117 *most geneticists still believed:* This notion, known as the "protein paradigm," is described, eg, in Kay 1993.

117 *Avery and his colleagues:* Avery, MacLeod, and McCarty 1944.

118 *"So there's the story":* The letter was written on May 13, 1943. It is part of "The Oswald T. Avery Collection," on the web under *Profiles in Science: National Library of Medicine,* at http://profiles.nlm.nih.gov/ps/retrieve/ResourceMetadata/CCBDVF.

118 *did not get the attention they deserved:* The fact that the paper was published in 1944, during the war, may have also contributed to its relatively low impact.

119 *"To my father, nucleic acids":* P. Pauling 1973.

119 *an unusual paper by biochemist:* Ronwin 1951.

119 *"in formulating a hypothetical structure":* Pauling and Schomaker 1952a.

119 *Ronwin retorted by pointing out:* He wrote to Pauling and directed him to a paper published by chemist Ludwig Anschütz in 1927, in which the latter suggested that phosphorus connected to five oxygen atoms in some structures.

119 *Pauling and Schomaker had to withdraw:* Pauling and Schomaker 1952b.

120 *Pauling heard that Maurice Wilkins:* Biochemist Gerald Oster wrote to Pauling about it on August 9, 1951. Oster interpreted Wilkins's delay in publishing the images as a lack of interest on his part, but Wilkins was, in fact, working toward confirming the results with better tools.

120 *Three separate events, all happening in 1951:* Although there are, of course, many accounts of the discovery of the structure of DNA, the autobiographical ones remain of special value, controversies notwithstanding. Watson 1980 (Norton Critical Edition) is particularly recommended. It includes, in addition to Watson's original (and controversial) text, an excellent selection of reviews and analyses. Also highly recommended are Crick 1988 and Wilkins 2003. Unfortunately, Rosalind Franklin did not live long enough to write her own autobiography, but two biographies—Sayre 1975 and Maddox 2002—fill that gap beautifully. Most recently, Franklin's sister, Jenifer Glynn, has written a wonderful memoir (Glynn 2012). Another interesting perspective on Franklin's experiences as a woman in a male-dominated lab is Des Jardins 2010, pp. 180–95.

121 *"molecular work on DNA in England":* Watson 1980, p. 13.

121 *"This means that as far as the experimental":* Randall wrote to Franklin on December 4, 1950. He added: "I am not in this way suggesting that we should give up all thought of work on solutions, but we do feel that the work on fibres would be more immediately profitable and, perhaps,

fundamental." The letter is reproduced, eg, in Olby 1974, p. 346, and in Maddox 2002, p. 114. See also Klug 1968a, b.

121 *"no doubt the brightest person":* Watson 1951, cited in Olby 1974, p. 354.

121 *"The interest was that his [Watson's] background":* Olby 1974, p. 310.

121 *"I have never seen Francis Crick":* Watson 1980, p. 9.

122 *"Jim was distinctly more outspoken":* Crick 1988, p. 64.

122 *"If either of us suggested":* Crick 1988, p. 70.

122 *"was determined to discover":* Crick 1988, p. 64.

123 *"it would not be fair to them":* John Randall wrote to Pauling on August 28, 1951. He started by explaining that contrary to Gerald Oster's interpretation, Wilkins was very interested in the DNA work: "I am sorry that Oster is rather misinformed about our intentions with regard to nucleic acid. Wilkins and others are busily engaged in working out the interpretation of the desoxyribosenucleic acid x-ray photographs." Pauling responded politely on September 25, 1951, saying that he was sorry for having troubled Randall. All the relevant documents are on the Oregon State University website.

123 *"One, thirty-five years old":* Chargaff 1978, p. 101.

123 *"So far as I could make out":* Chargaff 1978, p. 101. Chargaff added (p. 102): "I told them all I knew. If they had heard before about pairing rules, they concealed it."

123 *"You would be eccentric":* Crick in a recorded interview with Robert Olby; Olby 1974, p. 294.

123 *he and Crick produced their first model:* A draft describing their approach was written by Crick (Olby 1974, p. 357). Crick states clearly in the draft that his first model with Watson was "stimulated by the results presented by the workers at King's College, London, at a colloquium given on 21st November 1951." He also refers explicitly to Pauling's alpha-helix model.

124 *the reported water content was completely wrong:* Franklin found eight molecules per nucleotide, while Watson reported four molecules per lattice point.

124 *discovered some long-lost correspondence of Francis Crick:* Gann and Witkowski 2010.

125 *"I am afraid the average vote":* Gann and Witkowski 2010.

125 *"we've all agreed that we must come":* Gann and Witkowski 2010.

125 *she discovered that DNA occurred:* An excellent description of Franklin's work can be found in Klug 1968a, with some clarifications in Klug 1968b and additional information in Klug 1974. Elkin 2003 presents a historical perspective, and Braun, Tierney, and Schmitzer 2011 give a pedagogical explanation of the technical work.

126 *she absolutely refused to* assume: According to Klug 1968a, Franklin's antihelical attitude around May 1952 resulted from her uncertainty about the survivability of the helical structure in the A form of DNA. Her gen-

eral reluctance to *assume* anything about the structure is reflected in her statement "No attempt will be made to introduce hypotheses concerning details of structure at the present stage" (Franklin and Gosling 1953a).

127 *"He [Watson] just wanted the answer":* Crick 1988, p. 69.

127 *Elwyn Beighton in Astbury's lab:* Beighton's 1951 photograph of DNA is in the Special Collections, Astbury Papers, C7, at the University of Leeds. The photographs can be seen online at www.leeds.ac.uk/heritage/Astbury/Beighton_photo/index.html.

128 *Ruth B. Shipley, head of the Passport:* The entire episode is described in detail in Hager 1995, pp. 400–407. The general anticommunist atmosphere at the time is hauntingly depicted, eg, in Coute 1978.

129 *he immediately sent a letter to President:* Pauling wrote to Harry Truman on February 29, 1952.

129 *Pauling's passport trials and tribulations:* The *New York Times* ran a few stories and later discussed the whole passport system in relation to Pauling's problems on May 19, 1952, in an article entitled "Dr. Pauling's Predicament." The *Washington Post* wrote on May 13, 1952, "Pauling, Noted Chemist, Refused Passport," and Chicago's *Daily Sun-Times* ran on May 14, 1952, an article entitled "America's Own Iron Curtain."

130 *in the words of biologist Alex Rich:* Interview with the author on November 15, 2010.

130 *in that paper, published in 1950:* Chargaff 1950, and also Chargaff, Zamenhof, and Green 1950.

131 *Hershey and his collaborator, Martha Chase:* Hershey and Chase 1952.

132 *Williams showed amazingly detailed:* Williams 1952.

132 *the density measurements of Astbury and Bell:* An example of a diffraction photograph from DNA fibers obtained by Florence Bell in Astbury's lab can be seen in the University of Leeds' online collection at www.leeds.ac.uk/heritage/Astbury/Bell_Thesis/index.html; the published papers were Astbury and Bell 1938, and Astbury and Bell 1939.

132 *"The cylindrical molecule is formed":* Pauling and Corey 1953.

132 *"Because of their varied nature":* Pauling and Corey 1953.

133 *Pauling even had a small group:* Judson (1996, p. 131) heard about this meeting from a scientist who had worked at Caltech that winter. Pauling was apparently attempting to cheer himself up, given his political problems at the time.

134 *"We have, we believe, discovered":* Pauling's letter to Alexander Todd is on the Oregon State University website at http://osulibrary.orst.edu/specialcollections/coll/pauling/dna/corr/sci9.001.16-1p-todd-19521219.html.

134 *sent on the same day to Henry Allen Moe:* This letter is also on the Oregon State University website at http://osulibrary.orst.edu/specialcollections/coll/pauling/dna/corr/sci14.014.7-1p-moe-19521219.html.

134 *"To my left, near the window":* P. Pauling, 1973.

135 *"I was told a story today":* Letter from Peter Pauling to Linus, Ava Helen, and Crellin Pauling, Peter's brother. At http://osulibrary.orst.edu/ specialcollections/coll/pauling/dna/corr/bio5.041.6-peterpauling-paulings -19530113.html.

135 *"I wish Jim Watson were here":* Letter from Peter Pauling to Linus and Ava Helen Pauling. The transcript on the website reads mistakenly, "I am in direct manner," when it should be (see original) "in an indirect manner." At http://osulibrary.orst.edu/specialcollections/coll/pauling/dna/corr/ bio5.041.6-peterpauling-lp-19530123.html.

Chapter 7: Whose DNA Is It Anyway?

PAGE

136 *Watson rushed to Cambridge chemist:* Described in Watson 1980, p. 94.

137 *They went to celebrate at the Eagle:* Watson 1980, p. 95. Watson wrote, "Instead of sherry, I let Francis buy me a whiskey."

137 *"Pauling just didn't try":* Wilkins added, "He can't have looked closely at the details of what they did publish on base pairing, in that paper; almost all the details are simply wrong." Cited in Judson 1996, p. 80. Pauling him- self admitted to Judson (1996, p. 135), "We weren't working very hard on it."

139 *Clearly, this principle of self-complementarity:* Pauling himself mentioned this point in his second Hitchcock Foundation Lecture, "Chemical Bonds in Biology," given at the University of California at Berkeley, on January 17, 1983.

139 *When I talked to Alex Rich and Jack Dunitz:* I talked to Alex Rich on November 15, 2010, and to Jack Dunitz on November 23, 2010.

139 *Peter explained further that:* P. Pauling, 1973.

140 *Everyone engages in inductive reasoning:* In a series of seminal papers, Kahneman and Tversky discuss this topic in detail. See, eg, Kahneman and Tversky 1973, 1982. Also Kahneman, Slovic, and Tversky 1982, and Cos- mides and Tooby 1996. An excellent popular account is Kahneman 2011. Schulz 2010, pp. 115–32, discusses beautifully some aspects of inductive reasoning in relation to being wrong.

140 *"We can't live in a state":* Interview to *New Scientist*; Else 2011.

141 *He decided to gamble:* Lehrer 2009 gives a detailed description of the deci- sion process.

141 *cognitive bias known as the* framing effect: Kahneman 2011, pp. 363–74, gives many illuminating examples. Interestingly, fMRI studies show that the emotional responses in the amygdala (the brain region associated with negative feelings) in people who *realize* that "90 percent lean" is identical to "10 percent fat" are very similar to those in people who are actually

affected by the negative frame. The differences arise in the prefrontal cortex, which controls the emotions by thinking rationally about them. See, eg, de Martino et al. 2006.

142 *"Biologists probably consider"*: Letter from Linus Pauling to Henry Allen Moe, on December 19, 1952. At http://osulibrary.orst.edu/specialcollections/coll/pauling/dna; shcorr/sci14.014.7-lp-moe-19521219.html.

142 *"If that was such an important"*: This comment by Ava Helen was repeated by Pauling many times. See, eg, Hager 1995, p. 431.

142 *"The proposed structure accordingly permits"*: Pauling and Corey 1953, p. 96. This was an important point, since it shows that Pauling did relate the structure to information carrying capacity. Pauling and Corey also referred to the issue of amino acid sequencing, noting that in terms of the dimensions involved, nucleic acids are "well-suited to the ordering of amino-acid residues in a protein." This point was clearly made by Matt Meselson in his talk about Pauling. At http://osulibrary.oregonstate.edu/special collections/events/1995/paulingconference/video-s3-2-meselson.html.

142 *"I am just putting the final touches"*: Letter from Linus Pauling to Peter Pauling on March 27, 1953. At http://osulibrary.orst.edu/specialcollections/coll/pauling/dna/corr/sci9.001.33-lp-peterpauling19530327.html.

143 *Extensive studies by Swedish researchers:* In the context of the Betula Project (a research project on human memory), psychologist Lars-Göran Nilsson and his colleagues gave many memory tests to people in the age range thirty-five to eighty, and repeated tests at one-year intervals. The project started in 1988, and the researchers studied a total of 4,200 individuals. A collection of articles describing many of the results is Bäckman and Nyberg 2010.

143 *I asked molecular biologist:* Conversation on April 18, 2011. Jack Szostak, Nobel laureate in Physiology or Medicine, whom I also asked about Pauling's chemistry failure, also suggested that Pauling might have thought that he would find a way for the structure to work from a chemical perspective.

144 *"We have also been stimulated"*: Watson and Crick 1953a.

145 *The dark cross was the unmistakable:* Moreover, the spacing between successive dark spots indicates the distance covered by a complete turn of the helix (found to be 34 angstroms), and the distance between the center of the X shape (figure 14) to the top indicates the distance between successive bases.

145 *his "mouth fell open"*: Watson 1980, p. 98.

145 *Jerry Donohue came to the rescue:* At the time, there were uncertainties concerning the precise location of the hydrogen atoms in the bases. (There were different so-called tautomeric forms.) Donohue was an expert on the subject, eventually publishing important works in 1952 and 1955. His contribution to the successful DNA model was crucial.

145 *From the symmetry of the crystalline DNA:* In crystallography, symmetry (with respect to transformations such as rotation and reflection) is used to characterize the crystals. From the information in the report, Crick was able to deduce that the crystalline form of DNA could be described by what crystallographers call the "monoclinic C2" space group. This, in turn, implied that the chains were antiparallel. In an interview with Robert Olby, Crick admitted, "I don't think I would have thought of running them in the other direction" (Olby 1974, p. 404).

147 *"I enclose a draft of our letter":* Gann and Witkowski 2010.

148 *"If Rosy wants to see Pauling":* Letter of Wilkins to Crick, probably on March 23. Gann and Witkowski 2010.

148 *the landmark paper by Watson and Crick:* Watson and Crick 1953a.

149 *The details were presented in a second paper:* Watson and Crick 1953b.

150 *Crick explained later that this:* Crick 1988, p. 66.

150 *One was by Wilkins, Alexander Stokes, and Herbert Wilson:* Wilkins, Stokes, and Wilson 1953.

150 *The third paper in the April 25:* Franklin and Gosling 1953a. They published another paper in July of the same year, in which they detailed the distinction between the DNA A and B structures; Franklin and Gosling 1953b. See also Franklin and Gosling 1953c.

151 *became the title of a successful play:* For a review of the play *Photograph 51*, see, eg, http://theater.nytimes.com/2010/11/06/theater/06photograph.html.

151 *"It might be good for you":* Letter from Linus Pauling to Peter Pauling on March 27, 1953. At http://osulibrary.orst.edu/specialcollections/coll/pauling/dna/corr/sci9.001.33-lp-peterpauling-19530327.html.

152 *"Although it is only two months":* Pauling and Bragg 1953.

153 *"Failure hovers uncomfortably close":* Watson 2000.

154 *The new view that emerges:* See, eg, Reich et al. 2011, and interesting discussions on the blog page of paleoanthropologist John Hawks, *john hawks weblog.*

155 *Gamow was shown a copy:* Gamow's involvement and his coding schemes are described in detail, eg, in Judson 1996. Gamow also founded the RNA Tie Club, an organization that aspired, according to Gamow, "to solve the riddle of RNA structure, and to understand the way it builds proteins."

Chapter 8: *B* for Big Bang

PAGE

157 *"We now come to the question":* The entire event is described in detail in Mitton 2005, pp. 127–29. The program was announced in Britain's *Radio Times* magazine, March 28, 1949.

158 *The name has even survived:* Ferris 1993.

158 *Fred Hoyle was born:* Two excellent biographies of Hoyle are Mitton 2005 and Gregory 2005. Hoyle 1994 is a fascinating autobiography, as is the earlier and shorter Hoyle 1986a. Information can be found also through the Sir Fred Hoyle Project of St. John's College at the University of Cambridge. Online at www.joh.cam.ac.uk/library/special_collections/hoyle/project/#collection.

158 *"Between the ages of five":* Hoyle 1994, p. 42.

159 *In 1939 he decided to forego:* Hoyle wrote, "I discovered the Inland Revenue [the British equivalent of the IRS] distinguished between students and nonstudents by whether or not you had acquired the Ph.D."; Hoyle 1994, p. 127.

159 *"To achieve anything really worthwhile":* Hoyle 1994, p. 235. Hoyle added: "To hold popular opinion is cheap, costing nothing in reputation."

159 *"In 1926 it was possible":* Hoyle 1986b, p. 446.

160 *Dmitry Mendeleyev, a Russian chemist:* A number of other chemists came up with their own versions of the periodic table. The list included the French mineralogist Alexandre-Émile Béguyer de Chancourtois, John Newlands in England, and, in particular, Julius Lothar Meyer in Germany, who contributed similar tables (following some pioneering work by Robert Bunsen). Mendeleyev was the person, however, who managed to insert all sixty-two known elements into the table, and to not only predict elements awaiting discovery but also to even anticipate their densities and atomic weights. For a fascinating read on the periodic table, see Kean 2010.

161 *the smallest reproduction of the periodic table:* You can watch this feat on YouTube at www.geek.com/articles/geek-cetera/periodic-tablet-etched-on-a-single-hair-as-birthday-gift-20101230. See also *Science* 334, no. 7 (October 2011), p. 24.

161 *the English chemist William Prout:* For a brief biography of Prout (1785–1850), see Rosenfeld 2003.

161 *Eddington proposed in 1920:* Eddington 1920. At the time he still considered annihilation, too, as a possible source of energy. Eddington discussed the source of stellar energy in Eddington 1926.

161 *the French physicist Jean-Baptiste Perrin:* Wesemael 2009 described nicely the contributions of Perrin (1870–1942), and of the American physical chemist William Draper Harkins (1873–1951). See also Shaviv 2009, chapter 4.

162 *"go and find a hotter place":* Eddington 1926, p. 301.

162 *On one occasion, physicist Ludwik Silberstein:* The famous astrophysicist Subrahmanyan Chandrasekhar heard this story directly from Eddington. It is described in Berenstein 1973, p. 192.

164 *"Only the inertia of tradition":* Eddington 1920. Quoted also in full in the 1988 edition of *The Internal Constitution of the Stars* (Cambridge: Cambridge University Press), in the foreword (by S. Chandrasekhar), p. x.

165 *the strong, attractive nuclear force:* At distances that are very small com-

pared to the size of the nucleus, the nuclear force itself becomes repulsive, because particles such as protons (fermions) resist being crowded. This quantum effect is known as the Pauli exclusion principle.

165 *Using this quantum mechanical effect:* The probability of penetrating the barrier created by the Coulomb force increases exponentially with increasing energy of the particles. At the same time, the distribution of the particles at a given temperature is such that at high energies the number of particles decreases exponentially. The product of these two factors results in a peak (known as the Gamow peak) at which the nuclear reaction is most likely to occur. These ideas were first published in the late 1920s.

166 *In a remarkable paper published:* Bethe 1939.

166 *the proton-proton (p-p) chain:* For those with some nuclear physics background, the two main channels contributing to the energy production in the Sun are the pp I branch: $p + p \rightarrow D + e^+ + \nu_e$, $D + p \rightarrow {}^3He + \gamma$, ${}^3He + {}^3He \rightarrow {}^4He + 2p$, and the pp II branch: ${}^3He + {}^4He \rightarrow {}^7Be + \gamma$, ${}^7Be + e^- \rightarrow {}^7Li + \nu_e$, ${}^7Li + p \rightarrow 2\,{}^4He$.

166 *"There is no way in which nuclei":* Bethe 1939, p. 446.

167 *the results were published in the April 1:* Alpher, Bethe, and Gamow 1948. Gamow had already presented the idea of nucleosynthesis in the big bang in Gamow 1942 and Gamow 1946.

168 *often referred to as the "alphabetical article":* In his book *The Creation of the Universe*, Gamow jokes "There was, however, a rumor that later, when the α, β, γ theory went temporarily on the rocks, Dr. Bethe seriously considered changing his name to Zacharias" (Gamow 1961, p. 64).

168 *"stubbornly refuses to change his name":* Gamow 1961, p. 64.

168 *Even the great physicist Enrico Fermi:* Fermi examined the problem together with physicist Anthony Turkevich, even though they never published their results. A good description of the work on the mass gap problem can be found in Kragh 1996, pp. 128–32.

169 *was an epoch-making paper:* Hoyle 1946.

170 *"I sat in the RAS auditorium":* Hoyle made the presentation on November 8, 1946. Margaret Burbidge was at the time Margaret Peachey; she was to marry astronomer Geoffrey Burbidge in 1948. The quote is from a lecture Margaret Burbidge gave at St. John's College, Cambridge, on April 16, 2002. An excellent popular description of Hoyle's work on nucleosynthesis can be found in Mitton 2005, chapter 8.

170 *that particular student decided to ditch:* The incident is described in Hoyle 1986b.

171 *Ernst Öpik proposed in 1951:* Öpik 1951.

171 *Salpeter examined the triple alpha process:* Salpeter 1952. (Also Bondi and Salpeter 1952.) Salpeter went on to have a distinguished career in astrophysics.

172 *"Bad luck for poor old Ed"*: Hoyle 1982, p. 3.

172 "had to go a lot faster": Hoyle 1982, p. 3.

172 *Hoyle calculated that for carbon production:* While there had been some earlier suggestions for resonances around 7.4 MeV or so, those had never been confirmed, and in any case, no resonant level had been suggested (before Hoyle's prediction) above 7.5 MeV.

173 *What happened at that meeting:* After many years, the participants had somewhat different recollections of the events. A good summary of the various versions can be found in Kragh 2010.

173 *"Here was this funny little man"*: Interview by Charles Weiner, American Institute of Physics, in February 1973. Cited in Kragh 2010.

173 *"To my surprise, Willy didn't"*: Hoyle 1982, p. 3.

174 *Ward Whaling and his colleagues:* Described also in Fowler's Nobel lecture, "Experimental and Theoretical Nuclear Astrophysics; the Quest for the Origin of the Elements," given on December 8, 1983.

174 *In their just-over-one-page:* Dunbar, Pixley, Wenzel, and Whaling 1953. The paper and its significance is described also in Spear 2002.

174 *Despite his amazingly successful prediction:* Given that life as we know it is carbon based, much has been made of the anthropic significance of the resonant level in carbon. This issue is beyond the scope of the discussion here. I should note that in 1989, I, along with colleagues, showed that even if that energy level had been at a slightly different value, stars still would have produced carbon (Livio et al. 1989). This conclusion was confirmed later by more detailed work by Heinz Oberhummer and colleagues (Schlattl et al. 2004). For a detailed review, see Kragh 2010.

175 *more than a half year passed:* Hoyle et al. 1953.

175 *"In a sense this was but a minor"*: Hoyle 1986b, p. 449.

175 *"In the beginning God created"*: Gamow 1970, p. 127. Gamow really wanted to express his objections to the steady state theory (discussed in chapter 9) proposed by Hoyle, Bondi, and Gold, but he ended up nevertheless acknowledging Hoyle's contribution.

176 *"Perhaps his [Hoyle's] most important"*: Craafoord Prize 1997 press release.

176 *In this paper, published in 1954:* Hoyle 1954.

178 *The 1957 landmark paper:* Burbidge, Burbidge, Fowler, and Hoyle 1957. A very lively, popular account of the history of the theory of nucleosynthesis is Chown 2001. Tyson and Goldsmith 2004 provide a clear, humorous, multidisciplinary tour of cosmic evolution, from cosmology to biology.

180 *Both Fowler and Hoyle presented their results:* Hoyle 1958, p. 279; Fowler 1958, p. 269.

180 *a summary of the entire meeting:* Hoyle 1958, p. 431.

181 *felt that Hoyle should have also:* For an online discussion of the issue of Hoyle's not winning the Nobel Prize, see, eg, www.thelonggoodread

.com/2010/10/08/fred-hoyle-the-scientist-whose-rudeness-cost-him-a -nobel-prize.

181 *"The theory of stellar nucleosynthesis"*: Burbidge 2008. Nuclear astrophysicist Donald Clayton also explained the enormous significance of Hoyle's 1954 paper; Clayton 2007.

182 *"I also discovered that I had"*: Cited in Burbidge 2003, p. 218.

182 *and these collaborative exchanges*: Described beautifully in an interview with Tommy Gold by historian of science Spencer Weart. The interview took place on April 1, 1978, for the American Institute of Physics.

183 *Hoyle suggested cosmology as the topic*: Described in a fascinating interview with Fred Hoyle, in Lightman and Brawer 1990, p. 55.

Chapter 9: The Same Throughout Eternity?

PAGE

185 *"not only the laws of nature"*: Milne 1933.

186 *"In a sense, the steady-state"*: Hoyle 1990. In his excellent account of the history of the steady state theory, Kragh 1996 raised doubts about the authenticity of the film story. However, shortly after the *New York Times* reported (on May 24, 1952) on a lecture by the Astronomer Royal, Sir Harold Spencer Jones, Hoyle wrote him a letter in which he specifically mentioned the film story. The fact that this letter was written as early as 1952 gives this account more credibility.

187 *"What happened was that there was"*: Weart 1978.

189 *in a paper published in 1929*: Hubble 1929a.

190 *the Russian mathematician Aleksandr Friedmann*: Friedmann 1922.

191 *a passionate debate flared up*: A few of the articles about the credit for the discovery of cosmic expansion are Way and Nussbaumer 2011, Nussbaumer and Bieri 2011, Van den Bergh 2011, and Block 2011.

191 *astronomer Vesto Slipher had measured*: Described, eg, in Van den Bergh 1997.

191 *Arthur Eddington listed those*: Eddington 1923, p. 162.

191 *Georges Lemaître published (in French) a remarkable paper*: Lemaître 1927.

191 *from brightness measurements by Hubble*: Hubble 1926.

191 *Edwin Hubble obtained a value*: Hubble 1929a.

192 *Based solely on what I have described*: A brief summary of the events can be found in Livio 2011. See also Nussbaumer and Bieri 2009, Kragh and Smith 2003, and Trimble 2012 for more detailed descriptions.

192 *The English translation of Lemaître's*: Lemaître 1931a.

192 *Canadian astronomer Sidney van den Bergh*: Van den Bergh 2011.

193 *David Block went even somewhat further*: Block 2011.

193 *First, I obtained a copy*: I am grateful to the Archives Georges Lemaître

in Louvain, Belgium, and to Mme. Liliane Moens for providing me with a copy.

193 *any intent of extra editing:* Block thought that the "§§1–*n*" in the letter should be read as "§§1–72" because of the way the symbol "*n*" was written. He also interpreted the text as saying that Lemaître was given freedom to translate *only* the first seventy-two paragraphs of his paper. He further concluded that paragraph seventy-three was precisely Lemaître's equation determining the value of the Hubble constant. None of these was convincing. (See Livio 2011 for a discussion.)

194 *in the minutes of the council:* RAS 1931.

196 *The second piece of evidence:* RAS, RAS correspondence 1931.

198 *"The Expanding Universe":* Lemaître 1931b.

198 *"Dinner was a little late":* Bondi 1990, p. 191.

198 *Bondi and Gold proposed their Perfect Cosmological Principle:* Bondi and Gold 1948.

199 *Hoyle embarked on a more mathematical:* Hoyle 1948a.

200 *"causes unknown to science":* Hoyle 1948a.

200 *"Neutron creation appears to be the most likely":* Hoyle 1948a.

201 *"I shall not require of a scientific":* Popper 2006, p. 18.

202 *"Modern astrophysics appears to be":* Hoyle 1948b, p. 216.

203 *"Cosmology is one department of astronomy":* Greaves 1948, p. 216.

203 *"I am overawed by the whole":* Born 1948, p. 217.

204 *"inferior and detestable species":* Hoyle 1994, p. 270.

205 *one in the* New York Times: Appeared on May 24, 1952. The article in the *Christian Science Monitor* appeared on June 7, 1952.

206 *he and his student John Shakeshaft:* Described in the *Proceedings of Meeting of the Royal Astronomical Society* 886, pp. 104–6.

206 *"glad to see that there is now":* Gold 1955.

207 *Bondi was also skeptical:* Bondi 1955.

208 *"Was I being uncharitable in thinking":* Hoyle 1994, p. 410.

209 *The discovery of extremely active galaxies:* For an excellent popular description of the discovery of quasars, of the microwave background, and their significance see, eg, Rees 1997.

211 *could* all *still be explained:* Hoyle 1990.

212 *he published a book entitled:* Hoyle, Burbidge, and Narlikar 2000. Livio 2000 is a review of the book.

213 *Narlikar suggested that Hoyle's:* Interview with the author on March 5, 2012.

213 *Eggleton remembered Hoyle as a person:* Interview with the author on July 1, 2011.

214 *to describe the Victorian scholar Benjamin Jowett:* Jowett was appointed a fellow of Balliol College in Oxford at the age of twenty-one. He was satirized by:

First came I; my name is Jowett.
There's no knowledge, but I know it.
I am the Master of this college
What I don't know isn't knowledge.

214 *Faulkner admitted that he himself:* Interview with the author on August 19, 2011. See also Faulkner 2003.

214 *Martin Rees, Astronomer Royal:* Interview with the author on September 19, 2011. See also Rees 2001.

215 *"The problem with the scientific":* Hoyle 1994, p. 328.

216 *"From my very youth I despised":* Cited, eg, in Boorstin 1983, p. 345.

217 *Biologist Richard Dawkins labeled:* Hoyle's original argument was against abiogenesis—the theory for the origin of life on Earth—not against Darwin's theory of evolution. Dawkins expands on the discussion of Hoyle's fallacy in Dawkins 2006.

217 *Denial seldom evokes sympathy:* Kathryn Schulz gives a fascinating discussion of the sentiments involved in being wrong in Schulz 2010.

218 *Physicist Alan Guth proposed inflation:* He described the model beautifully in his popular book, Guth 1997.

218 *These are precisely the properties:* The relation between the steady state universe and the inflationary universe is discussed by Barrow 2005.

219 *contributed important studies to big bang:* In particular, Hoyle and Tayler 1964 and Wagoner, Fowler, and Hoyle 1967.

Chapter 10: The "Biggest Blunder"

PAGE

222 *Einstein himself first attempted:* Einstein 1917.

222 *Edwin Hubble confirmed unambiguously:* The definitive results were published in Hubble 1929b.

223 *"That term is necessary only":* Einstein 1917, p. 188 in the English translation.

223 *he modified his equations in such:* For the mathematically inclined, the original equations were: $G_{\mu\nu} = 8\pi G\, T_{\mu\nu}$, where G is the gravitational constant, $T_{\mu\nu}$ is the stress-energy tensor, and $G_{\mu\nu}$ is Einstein's curvature tensor representing the geometry of space-time. The modified equations were: $G_{\mu\nu} - 8\pi G\, \rho_\Lambda\, g_{\mu\nu} = 8\pi G\, T_{\mu\nu}$, where ρ_Λ could be taken as an energy density associated with the cosmological constant, and $g_{\mu\nu}$ is the space-time tensor that defines distances.

224 *Eddington was the first to point out:* Eddington 1930.

225 *Einstein insisted that the distribution of matter:* Einstein relied here on what is known as Mach's principle, after Austrian physicist and philosopher Ernst Mach, who suggested that motion and acceleration cannot be felt at all in an empty universe. An excellent discussion on the modern interpretation of Mach's principle can be found in Greene 2004.

225 *In his theory of special relativity:* There are many good popular books describing special and general relativity. Two that I found particularly engaging are Kaku 2004 and Galison 2003. Reading Einstein 2005 is always rewarding. In Tyson's witty 2007 collection of essays, he tackles many related topics beautifully.

226 *even for relative speeds as low:* Chou, Hume, Rosenband, and Wineland 2010.

226 *The theory was based largely:* Einstein himself explained the principles in Einstein 1955. Hawking 2007 presents a collection of Einstein's papers. In the scientific biography of Einstein, Pais 1982 explains the principles beautifully. Greene 2004 puts the theory in layperson's terms in the context of modern developments.

227 *"If a person falls freely":* The Kyoto Lecture was delivered on December 14, 1922. It was translated into English by Y. A. Ono, from notes taken by Yon Ishiwara (*Physics Today*, August 1932).

227 *the deviations recorded by his team:* The results were described in Dyson, Eddington, and Davidson 1920.

228 *Experiments have confirmed this effect:* New generations of clocks continuously improve the accuracy; eg, Tino et al. 2007.

229 *there were a few theoretical disappointments:* Earman 2001 gives a detailed, excellent (technical) discussion of Einstein's introduction of the cosmological constant and its early history. A clear exposition is also North 1965 (see also Norton 2000).

229 *Willem de Sitter found a solution:* de Sitter 1917.

230 *"If there is no quasi-static world":* Einstein's letter to Weyl on May 23, 1923.

231 *In a paper published in 1931:* Einstein 1931.

231 *in a paper Einstein published together:* Einstein and de Sitter 1932.

232 *In an article entitled "The Evolutionary Universe":* Gamow 1956.

232 *in his autobiographical book:* Gamow 1970, p. 44.

234 *"Einstein would meet me in his study":* Gamow 1970, p. 149.

235 *in his book* Ordinary Geniuses: Segrè 2011, p. 155.

235 *Albrecht Fölsing, who wrote one:* Fölsing 1997.

235 *he inquired with the army:* The entire episode is described in Brunauer 1986.

237 *Gamow asked for Einstein's opinion:* Letter written on September 24, 1946. Document 11-331 in the Albert Einstein Archives.

237 *Gamow attached his paper:* Letter written on July 9, 1948. Documents 11-333 and 11-334 in the Albert Einstein Archives.

237 *Einstein replied politely to Gamow's:* Eg., on August 4, 1948. Document 11-335 in the Albert Einstein Archives.

237 *Einstein's letter of August 4, 1946:* Document 70-960 in the Albert Einstein Archives.

237   *any other of his more intimate friends and colleagues:* The Physics Department at Princeton University held a symposium on relativity in honor of Einstein's seventieth birthday. Gamow was among the many who were invited. (A letter from Assistant to the Chairman at Princeton Paul Busse on March 15, 1949, informs him of travel arrangements.) However, Gamow's name does not appear on the list of people accepting the invitation, from March 17, 1949.

237   *"The introduction of the 'cosmological member'":* Einstein 1955, p. 127.

239   *"If Hubble's expansion had been discovered":* Einstein 1955, p. 127.

239   *included a supplementary footnote:* Pauli 1958, p. 220.

239   *"Our experience up to date":* Einstein 1934, p. 167.

240   *He articulated his feelings in a letter:* Letter written on September 26, 1947. Document 15-085.1 in the Albert Einstein Archives.

240   *This was a reply to a letter:* In his letter to Einstein on July 30, 1947, Lemaître says that he is making "some effort to modify" Einstein's attitude against the cosmological constant. Document 15-084.1 in the Albert Einstein Archives.

240   *"Since I have introduced this term":* Einstein's letter to Lemaître from September 26, 1947. Document 15-085.1 in the Albert Einstein Archives.

241   *Did he think then that this was:* Laloë and Pecker 1990 also did not think that Einstein had used this language, but the evidence they presented was much weaker.

241   *The laws of physics thus resemble:* This comparison was used also by Weinberg 2005.

241   *University of Manchester astronomer:* Leahy 2001.

242   *Einstein has become the embodiment:* Among the many biographies of Einstein, I want to mention in particular Isaacson 2007, Fölsing 1997, and a book that presents other aspects of his personality beautifully: Overbye 2000.

243   *Physicist Richard C. Tolman:* Letter on September 14, 1931. Document 23-031 in the Albert Einstein Archives.

243   *a universe with a cosmological constant:* Lemaître's ideas about galaxy formation were expressed, eg, in Lemaître 1931b, 1934.

243   *While this particular idea was shown:* Brecher and Silk 1969.

243   *"Return to the earlier view":* Eddington 1952, p. 24.

244   *"There are only two ways":* Eddington 1952, p. 25.

244   *The inflationary model:* Described beautifully in Guth 1997.

245   *he distinguished presciently between:* McCrea 1971.

Chapter 11: Out of Empty Space

PAGE

246   *Newton was the first to consider:* Calder and Lahav 2008 discuss how Newton's work alludes at least to some aspects of the effects of "dark energy."

246 *If one attempts to calculate:* Norton 1999 discusses this problem in detail.

247 *a few physicists attempted:* In particular, von Seeliger 1895 and Neumann 1896. Einstein may have been partially inspired by their work in introducing the cosmological constant.

247 *if the size of the universe somehow:* This model was suggested by Petrosian, Salpeter, and Szekeres 1967. However, a few years later, Petrosian showed that the model also predicted a decline in the brightness of more distant quasars, contrary to observations.

248 *When he introduced the cosmological constant:* Again, for the mathematically inclined, the new equation read: $G_{\mu\nu} - 8\pi G\, \rho_\Lambda\, g_{\mu\nu} = 8\pi G\, T_{\mu\nu}$, where ρ_Λ is the energy density associated with the cosmological constant.

248 *if one moves this term to the right-hand side:* The equation now reads: $G_{\mu\nu} = 8\pi G\, (T_{\mu\nu} + \rho_\Lambda\, g_{\mu\nu})$.

249 *This is an entirely different* physical: For excellent popular explanations of the cosmological constant as representing the energy of the vacuum see Krauss and Turner 2004, Randall 2011, and Greene 2011. Davies 2011 is also a short, accessible article. Theories of time, and their relations to cosmic expansion, are fascinatingly explained by Carroll 2001, and Frank 2011.

249 *Einstein proposed in 1919:* Einstein 1919.

249 *short note on the subject:* Einstein 1927.

249 *The practitioners of quantum mechanics:* Described in Enz and Thellung 1960.

250 *"Everything happens as though":* Lemaître 1934.

251 *Zeldovich made the first genuine:* Zeldovich 1967.

251 *when particle physicists carried out:* Excellent technical discussions of the cosmological constant problems can be found, eg, in Weinberg 1989, Peebles and Ratra 2003, and Carroll 2001 (updated regularly).

253 *two teams of astronomers:* The results were published by Riess et al. 1998 and Perlmutter et al. 1999. Overbye 1998 wrote a wonderful description of the discovery.

253 *The discovery of accelerating expansion:* Panek 2011, Kirshner 2002, Livio 2000, and Goldsmith 2000 provide colorful popular accounts of the discovery.

253 *Type 1a supernovae are very rare:* They are thought to result from white dwarfs that accrete mass up to the maximum mass allowed for a white dwarf (the Chandrasekhar mass). At that point, they ignite carbon at their centers. The entire white dwarf is destroyed in the explosion.

254 *combining detailed observations of the fluctuations:* The website of the Wilkinson Microwave Anisotrophy Probe (WMAP) provides updated information at www.map.gsfcnasa.gov.

255 *were hung on concepts such as supersymmetry:* Kane 2000 provides a beautiful popular description of the concepts involved in supersymmetry. Dine 2007 is an excellent technical text.

256   *The properties of our universe:* In the presentation here, I largely follow the discussion in Livio and Rees 2005. A classical book on anthropic reasoning is Barrow and Tipler 1986. Vilenkin 2006, Susskind 2006, and Greene 2011 give popular, comprehensive discussions of anthropics and the multiverse concept.

257   *physicist Steven Weinberg came up:* Weinberg 1987.

258   *who first presented this type:* Carter 1974.

259   *Wald was asked to examine data:* Mangel and Samaniego 1984 is a scholarly analysis of Wald's work on aircraft survivability. Wolfowitz 1952 chronicles all of Wald's work.

259   *very familiar with the* Malmquist bias: The Wikipedia article about the Malmquist bias is quite detailed and not too technical. At http://en.wikipedia.org/wiki/Malmquist_bias.

260   *Kepler published a treatise:* Kepler's model is described in some detail in Livio 2002, p. 142.

261   *This multiverse is supposed to continually:* Beautifully explained in Vilenkin 2006.

261   *in a vast cosmic landscape:* This "landscape" containing a huge number of potential universes is the subject of Susskind 2006.

263   *In a lecture delivered at Oxford:* Einstein 1934. The Herbert Spencer Lecture was delivered on June 10, 1933.

266   *Einstein's collaborator Leopold Infeld:* Infeld 1949, p. 477.

267   *"The history of science provides":* Lemaître 1949, p. 443.

267   *Einstein himself remained unconvinced:* Einstein 1949.

267   *Einstein's failures:* Weinberg 2005 presents a few of Einstein's mistakes. Ohanian 2008 gives an excellent compilation and review of all of Einstein's mistakes.

268   *"The aspiration to truth is more":* Einstein wrote his last autobiographical notes in March 1955, ending with comments about quantum mechanics. In Seelig 1956.

Coda

PAGE

270   *philosopher Bertrand Russell suggested:* Russell 1951.

270   *Psychologists Amos Tversky and Daniel Kahneman:* Kahneman 2011 gives a comprehensive, popular account on the ideas and findings about decision making.

271   *"We must, however, acknowledge":* Darwin 1998 [1874], p. 642.

BIBLIOGRAPHY

Alpher, R. A., Bethe, H., and Gamow, G. 1948. "The Origin of Chemical Elements." *Physical Review*, 73, 803.

Aristotle 4th century BCE. *The History of Animals*, book 9, chapter 6. Translation by D'Arcy Wentworth Thompson can be found at www.mlahanas.de/Greeks/Aristotle/HistoryOfAnimals9.html.

Armstrong, H. E. 1920. "Prof. John Perry, F. R. S." *Nature*, 105, 751.

Astbury, W. T. 1936. "X-Ray Studies of Protein Structure." *Nature*, 141, 803.

Astbury, W. T., and Bell, F. O. 1938. "Some Recent Developments in the X-Ray Study of Proteins and Related Structures." *Cold Spring Harbor Symposia on Quantitative Biology*, 6, 109.

———. 1939. "X-Ray Data on the Structure of Natural Fibres and Other Bodies of High Molecular Weight." *Tabulae Biologicae*, 17, 90.

Avery, D. T., MacLeod, C. M., and McCarty, M. 1944. "Studies on the Chemical Nature of the Substance Inducing Transformation of Pneumococcal Types: Induction of Transformation by a Desoxyribonucleic Acid Fraction Isolated from Pneumococcus Type III." *Journal of Experimental Medicine*, 79, 137.

Bäckman, L., and Nyberg, L. 2010. *Memory, Aging and the Brain: A Festschrift in Honour of Lars-Göran Nilsson* (Hove, UK: Psychology Press).

Barrow, J. D. 2005. "Worlds Without End or Beginnings." In *The Scientific Legacy of Fred Hoyle*. Edited by D. Gough (Cambridge: Cambridge University Press), 93.

Barrow, J. D., and Tipler, F. J. 1986. *The Anthropic Cosmological Principle* (Oxford: Clarendon Press).

Bechara, A., Damasio, H., and Damasio, A. R. 2000. "Emotion, Decision Making and the Orbitofrontal Cortex." *Cerebral Cortex*, 10, 295.

Becker, L. E. 1869. "On the Study of Science by Women." *Contemporary Review*, 10, January–April 1869, 389–90.

Becquerel, H. 1896. "Sur les Radiations invisibles émises par les corps phosphorescents." *Comptes Rendus de l'Académie des Sciences*, 122, 501.

Bell, G. 2008. *Selection: The Mechanism of Evolution,* 2nd ed. (Oxford: Oxford University Press).

Berenstein, J. 1973. *Einstein*, Modern Masters Series (New York: Viking).

Berridge, K. C. 2003. "Pleasures of the Brain." *Brain and Cognition*, 52, 106.

Bethe, H. A. 1939. "Energy Production in Stars." *Physical Review*, 55, 434.

Blackburn, H. 1902. *Women's Suffrage: A Record of the Women's Suffrage Movement in the British Isles* (London: Williams and Norgate).

Block, D. 2011. http://arxive.org/abs/1106.3928.

Bloom, P. 2010. *How Pleasure Works: The New Science of Why We Like What We Like* (New York: W. W. Norton).

Blow, D. 2002. *Outline of Crystallography for Biologists* (Oxford: Oxford University Press).

Bondi, H. 1955. "Proceedings at Meeting of the Royal Astronomical Society," No. 886, p. 106.

———. 1990. "The Cosmological Scene 1945–1952." In *Modern Cosmology in Retrospect*. Edited by B. Bertotti, R. Balbinot, S. Sergio, and A. Messina (Cambridge: Cambridge University Press).

Bondi, H., and Gold, T. 1948. "The Steady-State Theory of the Expanding Universe." *Monthly Notices of the Royal Astronomical Society*, 108, 252.

Bondi, H., and Salpeter, E. E. 1952. "Thermonuclear Reactions and Astrophysics." *Nature*, 169, 304.

Boorstin, D. J. 1983. *The Discoverers: A History of Man's Search to Know His World and Himself* (New York: Random House).

Born, M. 1948. In "Proceedings at Meeting of the Royal Astronomical Society," No. 847, p. 217.

Bowersox, J. 1999. "Experimental Staph Vaccine Broadly Protective in Animal Studies." *NIH News*, May 27, 1999.

Bowler, P. J. 2009. *Evolution: The History of an Idea, 25th Anniversary Edition* (Berkeley, CA: University of California Press).

Bozarth, M. A. 1994. "Pleasure Systems in the Brain." In *Pleasure: The Politics and the Reality*. Edited by D. M. Warburton (New York: John Wiley & Sons), 5.

Bragg, Sir W. L., Kendrew, J. C., and Perutz, M. F. 1950. "Polypeptide Chain Configurations in Crystalline Proteins." *Proceedings of the Royal Society of London*, A203, 321.

Brannigan, A. 1981. *The Social Basis of Scientific Discoveries* (Cambridge: Cambridge University Press).

Braun, G., Tierney, D., and Schmitzer, H. 2011. "How Rosalind Franklin Discovered the Helical Structure of DNA: Experiments in Diffraction." *Physics Teacher*, 49, 140.

Brecher, K., and Silk, J. 1969. "Lemaître Universe, Galaxy Formation and Observations." *Astrophysical Journal*, 158, 91.

Brehm, J. W. 1956. "Postdecision Changes in the Desirability of Alternatives." *Journal of Abnormal and Social Psychology*, 52(3), 384.

Brice, W. R. 1982. "Bishop Ussher, John Lightfoot and the Age of Creation." *Journal of Geological Education*, 30, 18.

Brownlie, A. D., and Lloyd Prichard, M. F. 1963. "Professor Fleeming Jenkin,

1833–1885, Pioneer in Engineering and Political Economy." *Oxford Economic Papers*, 15(3), 204.

Brunauer, S. 1986. "Einstein and the Navy: . . . An Unbeatable Combination." *On the Surface*. Naval Surface Weapons Center, January 24, 1986.

Bulmer, M. 2004. "Did Jenkin's Swamping Argument Invalidate Darwin's Theory of Natural Selection?" *British Journal for the History of Science*, 37(3): 281.

Burbidge, E. M., Burbidge, G. R., Fowler, W. A., and Hoyle, F, 1957. "Synthesis of the Elements in Stars," *Reviews of Modern Physics*, 29(4), 547.

Burbidge, G. 2003. "Sir Fred Hoyle." *Biographical Memoirs of Fellows of the Royal Society*, 49, 213.

———. 2008. "Hoyle's Role in B²FH," *Science*, 319, 1484.

Burchfield, J. D. 1990. *Lord Kelvin and the Age of the Earth* (Chicago: University of Chicago Press).

Burton, R. A. 2010. *On Being Certain: Believing You Are Right Even When You're Not* (New York: St. Martin's Griffin).

Calder, L., and Lahav, O. 2008. "Dark Energy: Back to Newton?" *Astronomy & Geophysics*, 49, 1.13.

Carozzi, A. V. 1969. *Telliamed, or Conversations between an Indian Philosopher and a French Missionary on the Diminution of the Sea* (Urbana, IL: University of Illinois Press).

Carroll, S. B. 2009. *Remarkable Creatures: Epic Adventures in the Search for the Origin of Species* (Boston: Houghton Mifflin Harcourt).

Carroll, S. B., Grenier, J. K., and Weatherbee, S. D. 2001. *From DNA to Diversity: Molecular Genetics and the Evolution of Animal Design* (Malden, MA: Blackwell Science).

Carroll, S. M. 2001. "The Cosmological Constant." *Living Reviews in Relativity*, 3, 1.

———. 2010. *From Eternity to Here: The Quest for the Ultimate Theory of Time* (New York: Dutton).

Carter, B. 1974. "Large Number Coincidences and the Anthropic Principle in Cosmology." In IAU Symposium 63, *Confrontation of Cosmological Theories with Observational Data* (Dordrecht: Reidel), 291.

Chabris, C., and Simons, D. 2010. *The Invisible Gorilla, and Other Ways Our Intuitions Deceive Us* (New York: Crown).

Chamberlin, T. C. 1899. "Lord Kelvin's Address on the Age of the Earth as an Abode Fitted for Life." *Science, New Series*, 9(235), 889.

Chapman, A. D. 2009. *Numbers of Living Species in Australia and the World*. 2nd ed. (Toowoomba, Australia: Australia Biodiversity Information Services).

Chargaff, E. 1950. "Chemical Specificity of Nucleic Acids and the Mechanism of their Enzymatic Degradation." *Experimentia*, 6, 201.

———. 1978. *Heraclitean Fire: Sketches from a Life before Nature* (New York: Rockefeller University Press).

Chargaff, E., Zamenhof, S., and Green, C. 1950. "Composition of Human Desoxypentose Nucleic Acid." *Nature*, 165, 756.

Chou, C. W., Hume, D. B., Rosenband, T., and Wineland, D. J. 2010. "Optical Clocks and Relativity," *Science*, 329, 1630.

Chown, M. 2001. *The Magic Furnace: The Search for the Origins of Atoms* (Oxford: Oxford University Press).

Cicero, M. T. 45 BCE. *The Nature of Gods*, p. 78; 1997. Translated with introduction and explanatory notes by P. G. Walsh (Oxford: Oxford University Press).

Clayton, D. D. 2007. "Hoyle's Equation." *Science*, 318, 1876.

Coleman, D. 1995. *Emotional Intelligence: Why It Can Matter More Than IQ* (New York: Bantam).

Cooper, J., and Fazio, R. H. 1984. "A New Look at Dissonance Theory." In *Advances in Experimental Social Psychology*. Edited by L. Berkowitz (New York: Academic Press).

Cosmides, L., and Tooby, J. 1996. "Are Humans Good Intuitive Statisticians After All? Rethinking Some Conclusions from Literature on Judgment Under Uncertainty." *Cognition*, 58, 1.

Coute, D. 1978. *The Great Fear: The Anti-Communist Purge Under Truman and Eisenhower* (New York: Touchstone).

Coyne, J. A. 2009. *Why Evolution Is True* (New York: Viking).

Coyne, J. A., and Orr, H. A. 2004. *Speciation* (Sunderland, MA: Sinauer).

Crick, F. 1988. *What Mad Pursuit: A Personal View of Scientific Discovery* (New York: Basic Books).

Curie, P., and Laborde, A. 1903. "Sur la chaleur dégagée spontanément par les sels de radium." *Comptes Rendus de l'Académie des Sciences*, 136, 673.

Dalrymple, G. B. 1991. *The Age of the Earth* (Stanford, CA: Stanford University Press).

———. 2001. "The Age of the Earth in the Twentieth Century: A Problem (Mostly) Solved." *Geological Society, London, Special Publications*, 190, 205.

Darwin, C. 1868. *The Variation of Animals and Plants Under Domestication* (London: John Murray).

———. 1909 [1842]. *The Foundations of the Origin of Species, A Sketch Written in 1842*. Edited by F. Darwin (Cambridge, Cambridge University Press).

———. 1958 [1892]. *The Autobiography of Charles Darwin and Selected Letters*. Edited by F. Darwin (New York: Dover Publications).

———. 1964 [1859]. *On the Origin of Species by Means of Natural Selection, or the Preservation of Favoured Races in the Struggle for Life* (London: John Murray). Reprinted (Cambridge, MA: Harvard University Press).

———. 1981 [1871]. *The Descent of Man, and Selection in Relation to Sex* (London: John Murray). Reprinted in facsimile with an introduction by J. T. Bonner and R. M. May (Princeton, NJ: Princeton University Press).

———. 1998. *The Descent of Man* (Amherst, NY: Prometheus Books). Originally published in the US 1874 (New York: Crowell).

———. 2009 [1859]. *The Annotated Origin: A Facsimile of the First Edition of On the Origin of Species*. Annotated by J. T. Costa (Cambridge, MA: Belknap Press of Harvard University Press).

Darwin, F. 1887. *The Life and Letters of Charles Darwin* (London: John Murray).

Darwin, F., and Seward, A. C. 1903. *More Letters of Charles Darwin: A Record of His Work in a Series of Hitherto Unpublished Letters* (New York: D. Appleton), Letter 406*, p. 36. Reprint 1972 (New York: Johnson).

Darwin, G. H. 1886. "Presidential Address to Section A." *BAAS Report*, 56, 511.

———. 1903. "Radio-Activity and the Age of the Sun." *Nature*, 68, 496.

———. 1907–16. In *The Scientific Papers of Sir George Darwin*. Edited by F. J. M. Stratton and J. Jackson. 5 vols. Reprinted 2009 (Cambridge: Cambridge University Press).

Davies, P. 2011. "Out of the Ether." *New Scientist*, 19, November, 50.

Davis, A. S. 1871. "The 'North British Review' and the Origin of Species." *Nature*, December 28, 161.

Dawkins, R. 1986. *The Blind Watchmaker* (New York: W. W. Norton).

———. 2006. *The God Delusion* (New York: Houghton Mifflin).

———. 2009. *The Greatest Show on Earth: The Evidence for Evolution* (New York: Free Press).

de Beer, G. 1964. "Mendel, Darwin, and Fisher." *Notes and Records of the Royal Society of London*, 19(2), 192.

Dein, S. 2001. "What Really Happens When Prophecy Fails: The Case of Lubavitch." *Sociology of Religion*, 62(3), 383.

de Maillet, B. 1748. *Telliamed ou entretiens d'un philosophe indien avec un missionaire françois sur la diminution de la mer, la formation de la Terre, l'origine de l'Homme etc.*, ed. J.-A. Guer (Amsterdam: L'Honoré et Fils). Translated and edited by Carozzi 1969.

de Martino, B., Kumaran, D., Seymour, B., and Dolan, R. J. 2006. "Frames, Biases, and Rational Decision-Making in the Human Brain." *Science*, 313, 684.

Dennett, D. C. 1995. *Darwin's Dangerous Idea: Evolution and the Meanings of Life* (New York: Simon & Schuster).

Depew, D. J., and Weber, B. H. 1995. *Darwinism Evolving: Systems Dynamics and the Genealogy of Natural Selection* (Cambridge, MA: MIT Press).

de Roode, J. 2007. "Reclaiming the Peppered Moth for Science." *New Scientist*, 8, December, 46.

de Sitter, W. 1917. "On the Relativity of Inertia: Remarks Concerning Einstein's Latest Hypothesis." *Proceedings of the Royal Academy of Amsterdam*, 19, 1217.

Des Jardins, J. 2010. *The Madame Curie Complex: The Hidden History of Women in Science* (New York: The Feminist Press).

Dine, M. 2007. *Supersymmetry and String Theory: Beyond the Standard Model* (Cambridge: Cambridge University Press).

Dobzhansky, T. 1973. "Nothing in Biology Makes Sense Except in the Light of Biology." *American Biology Teacher*, 35, 125.

Dover, G. 2000. *Dear Mr. Darwin: Letters on the Evolution of Life and Human Nature* (Berkeley, CA: University of California Press).

Dunbar, D. N. F., Pixley, R. E., Wenzel, W. A., and Whaling, W. 1953. "The 7.68-MeV State in C^{12}." *Physical Review*, 92, 649.

Dunitz, J. D. 1991. "Linus Pauling—Born 1901, Still Going Strong." *Croatica Chemica Acta*, 64(3), I.

Dyson, F. W., Eddington, A. S., and Davidson, C. 1920. "A Determination of the Deflection of Light by the Sun's Gravitational Field, from Observations Made at the Total Eclipse of May 29, 1919." *Philosophical Transactions of the Royal Society of London*, A 220, 291.

Earman, J. 2001. "Lambda: The Constant That Refuses to Die." *Archives for History of Exact Sciences*, 55, 190.

Eddington, A. S. 1920. "The Internal Constitution of the Stars." *Observatory*, 43, 341.

——. 1923. *The Mathematical Theory of Relativity* (Cambridge: Cambridge University Press).

——. 1926. *The Internal Constitution of the Stars* (Cambridge: Cambridge University Press).

——. 1930. "On the Instability of Einstein's Spherical World." *Monthly Notices of the Royal Astronomical Society*, 90, 668.

——. 1952. *The Expanding Universe* (Cambridge: Cambridge University Press).

Einstein, A. 1917. "Cosmological Considerations on the General Theory of Relativity." English translation of "Kosmologische Betrachtungen zur allgemeinen Relativitätstheorie," *Sitzungsberichte der Preussischen Akademie der Wissenschaften (PAW)*, 142.

——. 1919. In *PAW*, p. 249. Described also in Pais 1982, p. 287.

——. 1927. *The Formal Relationship of Riemann's Curvature Tensor to the Field Equilibria of Gravitation*, Mathematische Annalen, 97, 99.

——. 1931. In *PAW*, p. 235. Described also in Pais 1982, p. 288.

——. 1934. "On the Method of Theoretical Physics." *Philosophy of Science*, 1 (2), 163.

——. 1949. "Remarks Concerning the Essays Brought Together in this Co-operative Volume." In *Albert Einstein: Philosopher-Scientist*. Edited by P. A. Schilpp (Evanston, IL: Library of Living Philosophers).

——. 1955. *The Meaning of Relativity*, 5th ed. *Including the Relativistic Theory of the Non-Symmetric Field* (Princeton, NJ: Princeton University Press).

——. 1966. *The Meaning of Relativity*, 5th ed. *Including the Relativistic Theory of the Non-Symmetric Field* (Princeton, NJ: Princeton University Press).

———. 2005. *Relativity: The Special and General Theory.* Translated by R. W. Lawson, with introduction by R. Penrose, commentary by R. Geroch, historical essay by D. C. Cassidy (New York: Pi Press).

Einstein, A., and de Sitter, W. 1932. "On the Relation Between the Expansion and the Mean Density of the Universe." *Proceedings of the National Academy of Sciences,* 18(3), 213.

Elgvin, T. D., Hermansen, J. S., Fijarczyk, A., Bonnet, T., Borge, T., Saether, S. A., Voje, K. L., and Saetre, G.-P. 2011. "Hybrid Speciation in Sparrows II: A Role for Sex Chromosomes?" *Molecular Ecology,* 20(18), 3823.

Elkin, L. O. 2003. "Rosalind Franklin and the Double Helix." *Physics Today,* March, 42.

Else, L. 2011. "Nobel Psychologist Reveals the Error of Our Ways." *New Scientist* (magazine issue 2839), online at: 222.newscientist.com/article/mg21228390.400-nobel-psychologist-reveals-the-err-of-our-ways.html.

Endler, J. A. 1986. *Natural Selection in the Wild* (Princeton, NJ: Princeton University Press).

England, P., Molnar, P., and Richter, F. 2007. "John Perry's Neglected Critique of Kelvin's Age for the Earth: A Missed Opportunity in Geodynamics." *GSA Today* 17(1), 4.

Enz, C. P., and Thellung, A. 1960. "Nullpunktsenergie und Anordnung nicht vertauschbarer Faktoren im Hamiltonoperator." *Helvetica Physica Acta,* 33, 839.

Evans, L., and Smith, K. 1973. *Chess World Championship: Fischer vs. Spassky* (New York: Simon & Schuster).

Eve, A. S. 1939. *Rutherford: Being the Life and Letters of the Rt. Hon. Lord Rutherford, O. M.* (New York: Macmillan Company).

Faulkner, J. 2003. "Remembering Fred Hoyle." *Astrophysics and Space Science,* 285, 593.

Feller, S. A. 2010. "20th Century Physicists on Bank Notes." *Radiations,* 16(2), 7.

Ferris, T. 1993. "Needed: A Better Name for the Big Bang." *Sky & Telescope,* August 1993.

Festinger, L. 1957. *A Theory of Cognitive Dissonance* (Stanford, CA: Stanford University Press).

Fiorino, D. F., Coury, A., and Phillips, A. G. 1997. "Dynamic Changes in Nucleus Accumbens Dopamine Efflux During the Coolidge Effect in Male Rats." *Journal of Neuroscience,* 17(12), 4849.

Fisher, R. A. 1930. *The Genetical Theory of Natural Selection* (Oxford: Oxford University Press). A second edition was published in 1958 by Dover, New York.

Fölsing, A. 1997. *Albert Einstein: A Biography.* Translated by E. Osers (New York: Viking).

Foskett, D. J. 1953. "Wilberforce and Huxley on Evolution." *Nature*, 172, 920.

Fowler, W. A. 1958. "Nuclear Processes and Element Synthesis in Stars," in *Stellar Populations*. Edited by D. J. K. O'Connell, S. J. (Rome: Vatican Observatory).

Francoeur, E. 2001. "Molecular Models and the Articulation of Structural Constraints in Chemistry." In *Tools and Modes of Representation in Laboratory Science*. Edited by V. Klein (Dordrecht: Kluer).

Frank, A. 2011. *About Time: Cosmology and Culture at the Twilight of the Big Bang* (New York: Free Press).

Franklin, R. E., and Gosling, R. G. 1953a. "Molecular Configuration in Sodium Thymonucleate." *Nature*, 171, 740.

———. 1953b. "Evidence for a 2-Chain Helix in Crystalline Structure of Sodium Deoxyribonucleate." *Nature*, 172, 156.

———. 1953c. "The Structure of Sodium Thymonucleate Fibres. II: The Cylindrically Symmetrical Patterson Function." *Acta Crystallographica*, 6, 678.

Friedmann, D. 1922. "Über die Krümmung des Raumes." *Zeitschrift für Physik*, 10, 377.

Galison, P. 2003. *Einstein's Clocks, Poincaré's Maps: Empires of Time* (New York: W. W. Norton).

Gamow, G. 1942. "Concerning the Origin of Chemical Elements." *Journal of the Washington Academy of Sciences*, 32, 353.

———. 1946. "Expanding Universe and the Origin of Elements." *Physical Review*, 70, 572.

———. 1956. "The Evolutionary Universe." *Scientific American*, September, 136.

———. 1961. *The Creation of the Universe*, rev. ed. (New York: Viking).

———. 1970. *My World Line: An Informal Autobiography* (New York: Viking Press).

Gann, A., and Witkowski, J. 2010. "The Lost Correspondence of Francis Crick." *Nature*, 467, 419.

Gans, J., Wolinsky, M., and Dunbar, J. 2005. "Computational Improvements Reveal Great Bacterial Diversity and High Metal Toxicity in Soil." *Science*, 309, 1387.

Gess, R. W., Goates, M. I., and Rubidge, B. S. 2006. "A Lamprey from the Devonian Period of South Africa." *Nature*, 443, 981.

Glynn, J. 2012. *My Sister Rosalind Franklin* (Oxford: Oxford University Press).

Goertzel, T., and Goertzel, B. 1995. *Linus Pauling: A Life in Science and Politics* (New York: Basic Books).

Gold, T. 1955. "Proceedings at Meeting of the Royal Astronomical Society," No. 886, p. 106.

Goldsmith, D. 2000. *The Runaway Universe: The Race to Discover the Future of the Cosmos* (New York: Basic Books).

Gould, S. J. 2002. *The Structure of Evolutionary Theory* (Cambridge, MA: Belknap Press of Harvard University Press).

Gray, A. 1908. *Lord Kelvin: An Account of His Scientific Life and Work* (London: J. M. Dent and Company).

Greaves, W. M. H. 1948. In "Proceedings at Meeting of the Royal Astronomical Society," No. 847, p. 209.

Greene, B. 2004. *The Fabric of the Cosmos: Space, Time, and the Texture of Reality* (New York: Alfred A. Knopf).

———. 2011. *The Hidden Reality: Parallel Universes and the Deep Laws of the Cosmos* (New York: Alfred A. Knopf).

Gregory, T. 2005. *Fred Hoyle's Universe* (Oxford: Oxford University Press).

Guth, A. 1997. *The Inflationary Universe* (Reading, MA: Addison-Wesley).

Haber, F. C. 1959. *The Age of the Earth: Moses to Darwin* (Baltimore: Johns Hopkins Press).

Hager, T. 1995. *Force of Nature: The Life of Linus Pauling* (New York: Simon & Schuster).

Hardin, G. 1959. *Nature and Man's Fate* (New York: Signet).

Harrison, B. W. 2001. "Early Vatican Responses to Evolutionist Theology," at www.rtforum.org/it/it93.html.

Hartl, D. L., and Clark, A. G. 2006. *Principles of Population Genetics,* 4th ed. (Sunderland, MA: Sinauer Associates).

Hawking, S. 2007. *A Stubbornly Persistent Illusion: The Essential Scientific Writings of Albert Einstein* (Philadelphia: Running Press).

Henig, R. M. 2000. *The Monk in the Garden: The Lost and Found Genius of Gregor Mendel* (Boston: Houghton Mifflin).

Hershey, A. D., and Chase, M. 1952. "Independent Functions of Viral Proteins and Nucleic Acid in Growth of Bacteriophage." *Journal of General Physiology,* 36, 39.

Hodge, J., and Radick, G., eds. 2009. *The Cambridge Companion to Darwin* (Cambridge: Cambridge University Press).

Hodge, M. J. S. 1987. "Natural Selection as a Causal, Empirical, and Probabilistic Theory." In *The Probabilistic Revolution.* Edited by I. Krüger, G. Gigerenzer, and M. S. Morgan (Cambridge, MA: MIT Press), vol. 2, p. 233.

Holmes, A. 1947. "The Age of the Earth." *Endeavor,* 6, 99.

Hooper, J. 2003. *Of Moths and Men: An Evolutionary Tale* (New York: W. W. Norton).

Hoyle, F. 1946. "The Synthesis of the Elements from Hydrogen." *Monthly Notices of the Royal Astronomical Society,* 106, 343.

———. 1948a. "A New Model for the Expanding Universe." *Monthly Notices of the Royal Astronomical Society,* 108, 372.

———. 1948b. In "Proceedings at Meeting of the Royal Astronomical Society," No. 847, p. 209.

———. 1954. "On Nuclear Reactions Occurring in Very Hot Stars. I. The Synthesis of Elements from Carbon to Nickel." *Astrophysical Journal Supplement,* 1, 121.

——. 1958. "The Astrophysical Implications of Element Synthesis," in *Stellar Populations*. Edited by D. J. K. O'Connell, S. J. (Rome: Vatican Observatory).

——. 1982. "Two Decades of Collaboration with Willy Fowler." In *Essays in Nuclear Astrophysics: Presented to William A. Fowler on the Occasion of His Seventieth Birthday*. Edited by C. A. Barnes, D. D. Clayton, and D. N. Schramm (Cambridge: Cambridge University Press), p. 1.

——. 1983. *The Intelligent Universe* (New York: Holt, Rinehart and Winston).

——. 1986a. *The Small World of Fred Hoyle: An Autobiography* (London: Michael Joseph).

——. 1986b. "Personal Comments on the History of Nuclear Astrophysics." *Quarterly Journal of the Royal Astronomical Society*, 27, 445.

——. 1990. "An Assessment of the Evidence Against the Steady-State Theory." In *Modern Cosmology in Retrospect*. Edited by B. Bertotti, R. Balbinot, S. Bergio, and A. Messina (Cambridge: Cambridge University Press), 223.

——. 1994. *Home Is Where the Wind Blows: Chapters from a Cosmologist's Life* (Mill Valley, CA: University Science Books).

Hoyle, F., Burbidge, G., and Narlikar, J. V. 2000. *A Different Approach to Cosmology: From a Static Universe Through the Big Bang Towards Reality* (Cambridge: Cambridge University Press).

Hoyle, F., Dunbar, D. N. F., Wenzel, W. A., and Whaling, W. 1953. "A State in C^{12} Predicted from Astrophysical Evidence." *Physical Review*, 92, 1095.

Hoyle, F., and Tayler, R. J. 1964. "The Mystery of the Cosmic Helium Abundance." *Nature*, 203, 1108.

Hoyle, F., and Wickranasinghe, C. 1993. *Our Place in the Cosmos: The Unfinished Revolution* (London: J. M. Dent).

Hubble, E. P. 1926. "Extragalactic Nebulae." *Astrophysical Journal*, 64, 321.

——. 1929a. "A Relation Between Distance and Radial Velocity Among Extra-Galactic Nebulae." *Proceedings of the National Academy of Sciences USA*, 15, 168.

——. 1929b. "A Spiral Nebula as a Stellar System, Messier 31." *Astrophysical Journal*, 69, 103.

Hull, D. L. 1973. *Darwin and His Critics: The Reception of Darwin's Theory of Evolution by the Scientific Community* (Cambridge, MA: Harvard University Press).

Hutchinson, G. E. 1959. "Homage to Santa Rosalia; Or, Why Are There So Many Kinds of Animals?" *American Naturalist*, 93 (870), 145.

Hutton, J. 1788. "Theory of the Earth, or an Investigation of the Laws Observable in the Composition, Dissolution, and Restoration of Land upon the Globe." *Royal Society of Edinburgh Transactions*, 1, 209.

Huxley, T. H. 1909 [1869]. Originally in 1869, "Geological Reform," *Quarterly Journal of the Geological Society of London*, 25, 38–53; in 1909, *Discourses, Biological and Geological Essays* (New York: Appleton), p. 335.

Infeld, L. 1949. "On the Structure of Our Universe." In *Albert Einstein: Philosopher Scientist.* Edited by P. A. Schilpp (Evanston, IL: Library of Living Philosophers).

Isaacson, W. 2007. *Einstein: His Life and Universe* (New York: Simon & Schuster).

Jenkin, F. 1867. "Review of The Origin of Species," *North British Review,* June, vol. 46, 277.

Jensen, J. V. 1988. "Return to the Wilberforce-Huxley Debate." *British Journal for the History of Science,* 21(2), 161.

———. 1991. *Thomas Henry Huxley: Communicating for Science* (Newark, NJ: University of Delaware Press).

Joly, J. 1903. "Radium and the Geological Age of the Earth." *Nature,* 68, 526.

Judson, H. F. 1996. *The Eighth Day of Creation: Makers of the Revolution in Biology.* Expanded edition (Plainview, NY: Cold Spring Harbor Laboratory Press). Original edition 1979 (New York: Simon & Schuster).

Kahneman, D. 2011. *Thinking, Fast and Slow* (New York: Farrar, Straus and Giroux).

Kahneman, D., Slovic, P., and Tversky, A., eds. 1982. *Judgment Under Uncertainty: Heuristics and Biases* (Cambridge: Cambridge University Press).

Kahneman, D., and Tversky, A. 1973. "On the Psychology of Prediction." *Psychology Review,* 80, 237.

———. 1982. "On the Study of Statistical Intuition." *Cognition,* 11, 123.

Kaku, M. 2004. *Einstein's Cosmos: How Albert Einstein's Vision Transformed Our Understanding of Space and Time* (New York: W. W. Norton).

Kane, G. L. 2000. *Supersymmetry: Unveiling the Ultimate Laws of Nature* (New York: Basic Books).

Kant, I. 1754. "The Question, Whether the Earth Is Ageing, Considered Physically." Originally published (in German) in two parts in a Königsberg weekly. The English translation is in Reinhardt and Oldroyd 1982.

Kay, L. E. 1993. *The Molecular Vision of Life: Caltech, the Rockefeller Foundation, and the Rise of the New Biology* (New York: Oxford University Press).

Kean, S. 2010. *The Disappearing Spoon: And Other True Tales of Madness, Love, and the History of the World from the Periodic Table of the Elements* (New York: Little, Brown and Company).

Kelvin, Lord (Sir William Thomson). 1862. "On the Age of the Sun's Heat." *Macmillan's Magazine,* 5, 388. From reprint in *Popular Lectures and Addresses,* 1, 2nd ed., 356.

———. 1864. "On the Secular Cooling of the Earth." *Transactions of the Royal Society of Edinburgh,* 23, 167. From reprint in *Mathematical and Physical Papers,* 3, p. 295, 1890.

———. 1868. "On Geological Time," Address delivered before the Geological Society of Glasgow, February 27, 1868. *Popular Lectures and Addresses,* vol. 2, p. 10.

——. 1891–94. *Popular Lectures and Addresses*, 3 vols. (London: Macmillan and Co.).

——. 1895. "The Age of the Earth." *Nature*, 51, 438.

——. 1899. "The Age of the Earth as an Abode Fitted for Life." *Philosophical Magazine* (series 5), 47, 66.

——. 1904. "Contribution to the Discussion of the Nature of Emanations from Radium." *Philosophical* magazine, series 6, 7, 220.

Kelvin, Lord (Sir William Thomson), and Murray, J. R. 1895. "On the Temperature Variation of the Thermal Conductivity of Rocks." *Nature*, 52, 182.

Keynes, M. 2002. "Mendel — Both Ignored and Forgotten." *Journal of the Royal Society of Medicine*, 95(11), 576.

King, C. 1893. "The Age of the Earth." *American Journal of Science*, 45, 1.

Kirkaldy, J. F. 1971. *Geological Time* (Edinburgh: Oliver & Boyd).

Kirshner, R. 2002. *The Extravagant Universe: Exploding Stars, Dark Energy, and the Accelerating Cosmos* (Princeton, NJ: Princeton University Press).

Kirwan, R. 1797. "On the Primitive State of the Globe and Its Subsequent Catastrophe." *Transactions of the Royal Irish Society*, 6, 234.

Kitcher, P. 1982. *Abusing Science: The Case Against Creationism* (Cambridge, MA: MIT Press).

Kliman, R., Sheehy, B., and Schultz, J. 2008. "Genetic Drift and Effective Population Size." *Nature Education* 1(3).

Klug, A. 1968a. "Rosalind Franklin and the Discovery of the Structure of DNA." *Nature*, 219, 808.

——. 1968b. "Rosalind Franklin and DNA." *Nature*, 219, 880.

——. 1974. "Rosalind Franklin and the Double Helix." *Nature*, 248, 787.

Kragh, H. 1996. *Cosmology and Controversy: The Historical Development of Two Theories of the Universe* (Princeton, NJ: Princeton University Press), 173–74.

——. 2010. "An Anthropic Myth: Fred Hoyle's Carbon-12 Resonance Level." *Archive for History of Exact Sciences*, 64, 721.

Kragh, H., and Smith, R. W. 2003. "Who Discovered the Expanding Universe?" *History of Science*, 41, 141.

Krauss, L. M. 2012. *A Universe from Nothing: Why There Is Something Rather Than Nothing* (New York: Free Press).

Krauss, L. M., and Turner, M. S. 2004. "A Cosmic Conundrum." *Scientific American*, September 2004, 71.

Kritzman, L. D., ed., 2006. *The Columbia History of Twentieth-Century French Thought* (New York: Columbia University Press).

Kruger, J., and Dunning, D. 1999. "Unskilled and Unaware of It: How Difficulties in Recognizing One's Own Incompetence Lead to Inflated Self-Assessments." *Journal of Personality and Social Psychology*, 77(6), 1121.

Kunda, Z. 1990. "The Case for Motivated Reasoning." *Psychological Bulletin*, 108(3), 480.

Laloë, S., and Pecker, J.-C. 1990. "Where Did Einstein Lament Lambda?" *Physics Today*, 43(5), 117.

Leahy, J. P. 2001. "Einstein's Greatest Blunder: The Cosmological Constant," at www.jb.man.oc.uk/~jpl/cosmo/blunder.html.

Lee, S. W. S., and Schwartz, N. 2010. "Washing Away Postdecisional Dissonance." *Science*, 328(5979), 709.

Lehrer, J. 2009. *How We Decide* (Boston: Houghton Mifflin Harcourt).

Lemaître, G. 1927. "Un Univers homogène de masse constante et de rayon croissant, rendant compte de la vitesse radiale des nébuleuses extra-galactiques." *Annales de la Société Scientifique de Bruxelles*, A47, 49.

———. 1931a. "A Homogeneous Universe of Constant Mass and Increasing Radius Accounting for the Radial Velocity of Extra-Galactic Nebulae." *Monthly Notices of the Royal Astronomical Society*, 19, 483.

———. 1931b. "The Expanding Universe." *Monthly Notices of the Royal Astronomical Society*, 91, 490.

———. 1934. "Evolution of the Expanding Universe." *Proceedings of the National Academy of Sciences*, 20, 12.

———. 1949. "The Cosmological Constant." In *Albert Einstein: Philosopher Scientist*. Edited by P. A. Schilpp (Evanston, IL: Library of Living Philosophers).

Levene, P. A., and Bass, L. W. 1931. *Nucleic Acids* (New York: Chemical Catalog Company).

Lightman, A. 2005. *The Discoveries: Great Breakthroughs in 20th Century Science* (New York: Pantheon Books).

Lightman, A., and Brawer, R. 1990. *Origins: The Lives and Worlds of Modern Cosmologists* (Cambridge, MA: Harvard University Press).

Linden, D. J. 2011. *The Compass of Pleasure: How Our Brains Make Fatty Foods, Orgasm, Exercise, Marijuana, Generosity, Vodka, Learning, and Gambling Feel So Good* (New York: Viking).

Lindley, D. 2004. *Degrees Kelvin: A Tale of Genius, Invention, and Tragedy* (Washington, DC: Joseph Henry Press).

Livio, M. 2000. *The Accelerating Universe: Infinite Expansion, the Cosmological Constant, and the Beauty of the Cosmos* (New York: John Wiley & Sons).

———. 2000. "A Different Approach to Cosmology." *Physics Today*, 53, 71.

———. 2002. *The Golden Ratio: The Story of Phi, the World's Most Astonishing Number* (New York: Broadway Books).

———. 2011. "Lost in Translation: Mystery of the Missing Text Solved." *Nature*, 479, 171.

Livio, M., Hollowell, D., Weiss, A., and Truran, J. W. 1989. "The Anthropic Significance of the Existence of an Excited State of ^{12}C." *Nature*, 340, 281.

Livio, M., and Rees, M. J. 2005. "Anthropic Reasoning." *Science*, 309, 1022.

Lucas, J. R. 1979. "Wilberforce and Huxley: A Legendary Encounter." *Historical Journal*, 22, 313.

Lyell, C. 1830–33. *Principles of Geology Being an Attempt to Explain the Former Changes of the Earth's Surface, by Reference to Causes Now in Operation* (London: John Murray). Republished in 2009 (Cambridge: Cambridge University Press).

MacCurdy, E., ed. 1939. *The Notebooks of Leonardo da Vinci* (New York: G. Braziller).

Maddox, B. 2002. *Rosalind Franklin: The Dark Lady of DNA* (London: Harper Collins).

Majerus, M. E. N. 1998. *Melanism: Evolution in Action* (Oxford: Oxford University Press).

Mangel, M., and Samaniego, F. 1984. "Abraham Wald's Work on Aircraft Survivability." *Journal of the American Statistical Association*, 79, 259.

Marchant, J. 1916. *Alfred Russel Wallace: Letters and Reminiscences* (London: Cassell and Company).

Marinacci, B., ed. 1995. *Linus Pauling in His Own Words* (New York: Touchstone).

Mawer, S. 2006. *Gregor Mendel: Planting the Seeds of Genetics* (New York: Harry N. Abrams).

Mayr, E. 2001. *What Evolution Is* (New York: Basic Books).

McCrea, W. H. 1971. "The Cosmical Constant." *Quarterly Journal of the Royal Astronomical Society*, 12, 140.

McGrath, C. L., and Katz, L. A. 2004. "Genome Diversity in Microbial Eukaryotes." *Trends in Ecology and Evolution*, 19(1), 32.

McPherson, A. 2003. *Introduction to Macromolecular Crystallography* (Hoboken, NJ: John Wiley & Sons).

Mendel, G. 1866 [1865]. "Versuche über Pflanzen-Hybriden" ("Experiments in Plant Hybridization"), *Verhandlungen des naturforschenden Vereines Brünn*, 4, 3.

Meredith, R. W., et al. 2011. "Impacts of the Cretaceous Terrestrial Revolution and KPg Extinction on Mammal Diversification." *Science*, 334, 521.

Miller, D., ed. 1985. *Popper Selections* (Princeton: Princeton University Press).

Milne, E. A. 1933. "World-Structure and the Expansion of the Universe." *Zeitschrift für Astrophysik*, 6, 1.

Mirsky, A. E., and Pauling, L. 1936. "On the Structure of Native, Denatured, and Coagulated Proteins." *Proceedings of the National Academy of Sciences U.S.A.*, 22(7), 439.

Mitton, S. 2005. *Fred Hoyle: A Life in Science* (London: Aurum).

Moore, J. R. 1979. *The Post-Darwinian Controversies: A Study of the Protestant Struggle to Come to Terms with Darwin in Great Britain and America, 1870–1900* (Cambridge: Cambridge University Press).

Mora, C., Tittensor, D. P., Adl, S., Simpson, A. G. B., and Worm, B. 2011. "How Many Species Are There on Earth and in the Ocean?" *PLOS Biology* 9(8): e 1001127.doi:10.137i/journal.pbio.1001127.

Morris, S. W. 1994. "Fleeming Jenkin and the Origin of Species: A Reassessment." *British Journal for the History of Science*, 27, 313.

Motte, A. Translator. 1848. *Newton's Principia, with a Life of the Author by N. W. Chittenden* (New York: Daniel Adee).

Narasimhan, T. N. 2010. "Thermal Conductivity Through the 19th Century." *Physics Today*, August 2010, 36.

Nernst, W. 1916. "Über einen Versuch, von quantentheoretischen Betrachtungen zur Annahme stetiger Energieänderungen surückzukehren." *Verhandlungen der Deutschen Physikalischen Gesellschaft*, 18, 83.

Nestler, E. J., and Malenka, R. C. 2004. "The Addicted Brain." *Scientific American*, March, 78.

Neumann, C. 1896. *Allgemeine Untersuchungen über das Newton'sche Princip der Fernwirkungen, mit besonderer Rücksicht auf die elektrischen Wirkungen* (Leipzig: Teubner).

Newton, I. 1687. *Philosophiae Naturalis Principia Mathematica* (London: S. Pepys, Royal Society Press).

North, J. D. 1965. *The Measure of the Universe: A History of Modern Cosmology* (Oxford: Clarendon Press).

Norton, J. D. 1999. "The Cosmological Woes of Newtonian Gravitation Theory." In *The Expanding Worlds of General Relativity: Einstein Studies*. Edited by H. Goenner, J. Renn, J. Ritter, and T. Sauer (Boston: Birkhaüser), 7, 271.

———. 2000. "Nature Is the Realisation of the Simplest Conceivable Mathematical Ideas: Einstein and the Canon of Mathematical Simplicity." *Studies in History and Philosophy of Modern Physics*, 31(2), 135.

Nudds, J. R., McMillan, N. D., Weaire, D. C., and McKenna Lawlor, S. M. P., eds. 1988. *Science in Ireland, 1800–1930: Tradition and Reform* (Dublin: privately published, Trinity College).

Nussbaumer, H., and Bieri, L. 2009. *Discovering the Expanding Universe* (Cambridge: Cambridge University Press).

———. 2011. http://arxiv.org/abs/1107.2281.

Nye, M. J. 2001. "Paper Tools and Molecular Architecture in the Chemistry of Linus Pauling." In *Tools and Modes of Representation in Laboratory Sciences*. Edited by V. Klein (Dordrecht: Kluwer).

Ochs, V. L. 2005. "Waiting for the Messiah, a Tambourine in Her Hand." *Nashim: A Journal of Jewish Women's Studies & Gender Issues*, (9), 144.

Ohanian, H. C. 2008. *Einstein's Mistakes: The Human Failings of Genius* (New York: W. W. Norton & Company).

Olby, R. 1974. *The Path to the Double Helix* (London: Macmillan).

Olds, J. 1956. "Pleasure Centers in the Brain." *Scientific American*, October, 105.

Olds, J., and Milner P. 1954. "Positive Reinforcement Produced by Electrical Stimulation of Septal Area and Other Regions of Rat Brain." *Journal of Comparative and Physiological Psychology*, 47, 419.

Öpik, E. 1951. "Stellar Models with Variable Composition. II: Sequences of Models with Energy Generation Proportional to the Fifteenth Power of Temperature." *Proceedings of the Royal Irish Academy,* A 54, 49.

Orel, V. 1996. *Gregor Mendel: The First Geneticist.* Translated by S. Finn (New York: Oxford University Press).

Overbye, D. 1998. "A Famous Einstein 'Fudge' Returns to Haunt Cosmology." *New York Times,* May 26, 1998.

———. 2000. *Einstein in Love: A Scientific Romance* (New York: Viking).

Pais, A. 1982. *Subtle Is the Lord: The Science and Life of Albert Einstein* (Oxford: Oxford University Press).

Paley, W. 1802. *Natural Theology, or Evidence of the Existence and Attributes of the Deity, Collected from the Appearances of Nature.* 2006. Edited with an introduction and notes by M. D. Eddy and D. Knight (Oxford: Oxford University Press).

Pallen, M. 2009. *The Rough Guide to Evolution* (London: Rough Guides).

Panek, R. 2011. *The 4% Universe: Dark Matter, Dark Energy, and the Race to Discover the Rest of Reality* (Boston: Houghton Mifflin Harcourt).

Parshall, K. H. 1982. "Varieties As Incipient Species: Darwin's Numerical Analyses." *Journal of the History of Biology,* 15(2), 191.

Patterson, C. 1956. "Age of Meteorites and the Earth." *Geochimica et Cosmochimica Acta,* 10(4), 230.

Pauli, W. 1958. *Theory of Relativity.* Translated by G. Field (Oxford: Pergamon Press). Reprinted 1981 (Mineola, NY: Dover).

Pauling, L. 1935. "The Oxygen Equilibrium of Hemoglobin and Its Structural Interpretation." *Science,* 81, 421.

———. 1939. *The Nature of the Chemical Bond and the Structure of Molecules and Crystals* (Ithaca, NY: Cornell University Press).

———. 1948a. "Nature of Forces Between Large Molecules of Biological Interest." *Nature,* 161, 707.

———. 1948b. "Molecular Architecture and the Processes of Life." 21st Sir Jesse Boot Foundation Lecture, Nottingham, England. Lecture given on May 28, 1948.

———. 1955. "The Stochastic Method and the Structure of Proteins." *American Scientist,* 43, 285.

———. 1996. "The Discovery of the Alpha Helix." *Chemical Intelligencer,* January, 32 (published by Dorothy Munro).

Pauling, L., and Bragg, L. 1953. "Discussion des Rapports de MM L. Pauling et L. Bragg." *Rep. Institut International de Chimie Solvay,* 111.

Pauling, L., and Corey, R. B. 1950. "Two Hydrogen-Bonded Spiral Configurations of the Polypeptide Chain." *Journal of the American Chemical Society,* 72(11), 5349.

———. 1953. "A Proposed Structure for the Nucleic Acids." *Proceedings of the National Academy of Sciences U.S.A.,* 39, 84.

Pauling, L., Corey, R. B., and Branson, H. R. 1951. "The Structure of Proteins: Two Hydrogen-Bonded Helical Configurations of the Polypeptide Chain." *Proceedings of the National Academy of Sciences U.S.A.*, 37, 205.

Pauling, L., and Coryell, C. D. 1936. "The Magnetic Properties and Structure of Hemoglobin and Carbonmonoxyhemoglobin." *Proceedings of the National Academy of Sciences*, 22, 210.

Pauling, L., and Schomaker, V. 1952a. "On a Phospho-tri-anhydride Formula for the Nucleic Acids." *Journal of the American Chemical Society*, 74, 1111

———. 1952b. "On a Phospho-tri-anhydride Formula for the Nucleic Acids." *Journal of the American Chemical Society*, 74, 3712.

Pauling, P. 1973. "DNA—The Race That Never Was?" *New Scientist*, May 31, 558.

Peckham, M., ed. 1959. *The Origin of Species: A Variorum Text* (Philadelphia: University of Pennsylvania Press).

Peebles, P. J. E., and Ratra, B. 2003. "The Cosmological Constant and Dark Energy." *Review of Modern Physics*, 75, 559.

Perlmutter, S., et al. 1999. *Astrophysical Journal*, 517, 565.

Perry, J. 1895a. "On the Age of the Earth." *Nature*, 51, 224.

———. 1895b. "On the Age of the Earth." *Nature*, 51, 341.

———. 1895c. "The Age of the Earth." *Nature*, 51, 582.

Perutz, F. 1987. "I Wish I'd Made You Angry Earlier." *Scientist*, 1(7), 19.

Petrosian, V., Salpeter, E., and Szekeres, P. 1967. "Quasi-Stellar Objects in the Universe with Non-Zero Cosmological Constant." *Astrophysical Journal*, 147, 1222.

Philo of Alexandria 1st century CE. *Allegories of the Sacred Laws*. Cited in Toumlin and Goodfield 1965, p. 58. The treatise is online at www.early christianwritings.com/yonge/book2.html.

Pliny, the Elder 1st century CE. *The Natural History*, book 8, chapter 37. Edited by J. Bostock and H. T. Riley (London: Taylor & Francis, 1855).

Popper, K. 1976. *Unended Quest: An Intellectual Autobiography* (Glasgow: Fontana/Collins).

———. 1978. "Natural Selection and the Emergence of Mind." *Dialectica*, 32, 339.

———. 2006. *The Logic of Scientific Discovery* (London: Routledge). First published 1935, *Logik der Forschung* (Vienna: Verlag von Julius Springer).

Randall, L. 2011. *Knocking on Heaven's Door: How Physics and Scientific Thinking Illuminate the Universe and the Modern World* (New York: Ecco).

RAS 1931. Royal Astronomical Society Papers 2. *Minutes of Council*, 12, 160, 165, 166.

Rees, M. 1997. *Before the Beginning: Our Universe and Others* (Reading, MA: Helix Books).

———. 2001. "Fred Hoyle." *Physics Today*, November 2001, 75.

Reich, D., Patterson, N., Kircher, M., et al. 2011. "Denisova Admixture and the First Modern Human Dispersals into Southeast Asia and Oceania." *American Journal of Human Genetics*, 89, 516.

Reinhardt, O., and Oldroyd, D. R. 1982. "Kant's Thoughts on the Ageing of the Earth." *Annals of Science*, 39, 349.

Richter, F. M. 1986. "Kelvin and the Age of the Earth." *Journal of Geology*, 94, 395.

Ridley, M. 2004a. *Evolution*, 3rd ed. (Malden, MA: Blackwell Science).

——, ed. 2004b. *Evolution*, 2nd ed. (Oxford: Oxford University Press).

Riess, A. G., et al. 1998. *Astronomical Journal*, 116, 1009.

Ronwin, E. 1951. "A Phospho-tri-anhydride Formula for the Nucleic Acids." *Journal of the American Chemical Society*, 73, 5141.

Rose, M. R. 1998. *Darwin's Spectre: Evolutionary Biology in the Modern World* (Princeton, NJ: Princeton University Press).

Rosenfeld, L. 2003. "William Prout: Early 19th Century Physician-Chemist." *Clinical Chemistry*, 49(4), 699.

Ruse, M., and Richards, R. J., eds. 2009. *The Cambridge Companion to the "Origin of Species"* (Cambridge: Cambridge University Press).

Russell, B. 1951. "The Answer to Fanaticism: Liberalism." In the *New York Times Magazine*, December, 16, 1951.

Salisbury, R. C. 1894. President's address, *Report of the British Association for the Advancement of Science*, Oxford, p. 3.

Salpeter, E. E. 1952. "Nuclear Reactions in Stars Without Hydrogen." *Astrophysical Journal*, 115, 326.

Sayre, A. 1975. *Rosalind Franklin and DNA* (New York: W. W. Norton).

Schilthuizen, M. 2001. *Frogs, Flies, and Dandelions: The Making of a Species* (Oxford: Oxford University Press).

Schlattl, H., Heger, A., Oberhummer, H., Rauscher, T., and Csóto, A. 2004. "Sensitivity of the C and O Production on the 3α Rate." *Astrophysics and Space Science*, 291, 27.

Schulz, K. 2010. *Being Wrong: Adventures in the Margin of Error* (New York: HarperCollins).

Sclater, A. 2003. "The Extent of Charles Darwin's Knowledge of Mendel." *Georgia Journal of Science*, 61, 134.

Seelig, C., ed. 1956. *Helle Zeit – Dunkle Zeit* (Zürich: Europa Verlag).

Segrè, G. 2011. *Ordinary Geniuses: Max Delbruck, George Gamow, and the Origins of Genomics and Big Bang Cosmology* (New York: Viking).

Serafini, A. 1989. *Linus Pauling: A Man and His Science* (New York: Paragon House).

Sharlin, H. I., and Sharlin, T. 1979. *Lord Kelvin: The Dynamic Victorian* (University Park, PA: Penn State University Press).

Shaviv, G. 2009. *The Life of Stars: The Controversial Inception and Emergence of the Theory of Stellar Structure* (Heidelberg: Springer).

Shipley, B. C. 2001. " 'Had Lord Kelvin a Right?': John Perry, Natural Selection and the Age of the Earth, 1894–1895." In *The Age of the Earth: From 4004 BC to AD 2002*. Edited by C. L. E. Lewis and S. J. Knell, Geological Society, London, Special Publications, 190, 91.

Sidgwick, I. 1898. "A Grandmother's Tales." *Macmillan's Magazine*, 78(1), 433.

Smith, C., and Wise, M. N. 1989. *Energy and Empire: A Biographical Study of Lord Kelvin* (Cambridge: Cambridge University Press).

Soddy, F. 1904. *Radio-Activity: An Elementary Treatise from the Standpoint of Disintegration Theory* (London: The Electrician).

——. 1906. "The Recent Controversy on Radium." *Nature*, 74, 516.

Spear, R. 2002. "The Most Important Experiment Ever Performed by an Australian Physicist." *Physicist*, 39(2), 35.

Spinoza, B. 1925. *Spinoza Opera*. Edited by C. Gebhardt (Heidelberg: Carl Winter).

Stacey, F. D. 2000. "Kelvin's Age of the Earth Paradox Revisited." *Journal of Geophysical Research*, 105 (B6), 13, 155.

Sturchio, N. C., and Purtschert, R. 2012. "Kr-81 Case Study: The Nubian Aquifer (Egypt)." In *Dating Old Groundwater: A Guide Book*. Edited by A. Suckow (Vienna: IAEA).

Susskind, L. 2006. *The Cosmic Landscape: String Theory and the Illusion of Intelligent Design* (New York: Little, Brown and Company).

Tait, G. G. 1869. "Geological Time." *North British Review*, July, 406.

Taylor, A. J. P. 1963. "Mistaken Lessons from the Past." *Listener*, June 6.

Thompson, S. P. 1910. *The Life of William Thomson, Baron Kelvin of Largs* (London: Macmillan and Co.). Reprinted 1976 (New York: Chelsea Publishing Company).

Thomson, J. J. 1936. *Recollections and Reflections* (London: Bell).

Tino, G. M., et al. 2007. "Atom Interferometers and Optical Atomic Clocks: New Quantum Sensors for Fundamental Physics Experiments in Space." *Nuclear Physics B* (Proceedings Supplements), 166, 159.

Toumlin, S. E., and Goodfield, J. 1965. *The Discovery of Time* (New York: Harper & Row).

Trimble, V. 2012. "Eponyms, Hubble's Law, and the Three Princes of Parallax." *Observatory*, 132, 33.

Tyson, N. d-G. 2007. *Death by Black Hole: And Other Cosmic Quandaries* (New York: W. W. Norton).

Tyson, N. dG., and Goldsmith, D. 2004. *Origins: Fourteen Billion Years of Cosmic Evolution* (New York: W. W. Norton).

Van den Bergh, S. 1997. In *The Extragalactic Distance Scale*. Edited by M. Livio, M. Donahue, and N. Panagia (Cambridge: Cambridge University Press), p. 1.

——. 2011. http://arxiv.org/abs/1106.1195.

Van Overwalle, F., and Jordens, K. 2002. "An Adaptive Connectionist Model of Cognitive Dissonance." *Personality and Social Psychology Review*, 6(3), 204.

Van Veen, V., Krug, M. K., Schooler, J. W., and Carter, C. S. 2009. "Neural Activity Predicts Attitude Change in Cognitive Dissonance." *Nature Neuroscience*, 12(11), 1469.

Vila, R., Bell, C. D., Macniven, R., Goldman-Huertas, B., Ree, R. H., Marshall, C. R., Balient, S., Johnson, K., Benjamini, D., and Pierce, N. 2011. "Phylogeny and Palaeoecology of *Polyommatus* Blue Butterflies Show Beringia Was a Climate-Regulated Gateway to the New World." *Proceedings of the Royal Society*, series B, 278.

Vilenkin, A. 2006. *Many Worlds in One: The Search for Other Universes* (New York: Hill and Wang).

von Seeliger, H. 1895. "Über das Newton'sche Gravitationsgesetz." *Astronomische Nachrichten* 137, 129.

Vorzimmer, P. 1963. "Charles Darwin and Blending Inheritance." *Isis*, 54(3), 371.

Wagoner, R. V., Fowler, W. A., and Hoyle, F. 1967. "On the Synthesis of Elements at Very High Temperatures." *Astrophysical Journal*, 148, 3.

Watson, J. D. 1951. Letter to biophysicist Max Delbrück, dated December 9, 1951, Caltech Archives.

———. 1980. *The Double Helix: A Personal Account of the Discovery of the Structure of DNA*. Edited by G. S. Stent. A Norton Critical Edition (New York: W. W. Norton).

———. 2000. *A Passion for DNA: Genes, Genomes, and Society* (Oxford: Oxford University Press), 44.

Watson, J. D., and Crick, F. H. C. 1953a. "Molecular Structure of Nucleic Acids." *Nature*, 171, 737.

———. 1953b. "Genetical Implications of the Structure of Deoxyribonucleic Acid." *Nature*, 171, 964.

Way, M., and Nussbaumer, H. 2011. "Lemaître's Hubble Relationship." *Physics Today*, August 2011, 8.

Weart, S. 1978. "Oral History Transcript—Dr. Thomas Gold." *Source for History of Modern Astrophysics*. Niels Bohr Library & Archives (College Park, MD: American Institute of Physics), 34.

Weinberg, S. 1987. "Anthropic Bound on the Cosmological Constant." *Physical Review Letters*, 59, 2607.

———. 1989. "The Cosmological Constant Problem." *Review of Modern Physics*, 61(1), 1.

———. 1992. *Dreams of a Final Theory* (New York: Pantheon).

———. 2005. "Einstein's Mistakes." *Physics Today*, 58(11), 31.

Wells, J. 2000. *Icons of Evolution: Science or Myth?* (Washington, DC: Regency Publishing).

Wesemael, F. 2009. "Harkins, Perrin and the Alternative Paths to the Solution of the Stellar-Energy Problem, 1915–1923," *Journal for the History of Astronomy*, 40, No. 3, 277.

Westen, D., Blagov, P. S., Horenski, K., Kelts, C., and Hamman, S. 2006. "Neural Bases of Motivated Reasoning: An fMRI Study of Emotional Constraints on Partisan Political Judgment in the 2004 US. Presidential Election." *Journal of Cognitive Neuroscience*, 18(11), 1947.

Wilkins, M. 2003. *The Third Man of the Double Helix: The Autobiography of Maurice Wilkins* (Oxford: Oxford University Press).

Wilkins, M. H. F., Stokes, A. R., and Wilson, H. R. 1953. "Molecular Structure of Deoxypentase Nucleic Acids." *Nature*, 171, 738.

Williams, R. C. 1952. "Electron Microscopy of Sodium Desoxyribonucleate by Use of a New Freeze-Drying Method." *Biochimica et Biophysica Acta*, 9, 237.

Wilson, D. B. 1987. *Kelvin and Stokes: A Comparative Study in Victorian Physics* (Bristol: Adam Hilger).

Wilson, E. B. 1925. *The Cell in Development and Heredity*, 3rd ed. (New York: Macmillan).

Wilson, E. O. 1992. *The Diversity of Life* (Cambridge, MA: Belknap Press).

Wilson, J. D. 1999. "Watson on Pauling." *Time* magazine, March 21, 1999. Online at www.time.com/time/magazine/article/0,9171, 21848,00.html.

Wilson, W. 1913. *The New Freedom: A Call for the Emancipation of the Generous Energies of a People* (New York: Doubleday), chapter 2.

Wilson, W. E. 1903. "Radium and Solar Energy." *Nature*, 68, 222.

Wise, R. A. 1998. "Drug-Activation of Brain Reward Pathways." *Drug and Alcohol Dependence*, 51(1–2), 13.

Wolfowitz, J. 1952. "Abraham Wald 1902–1950." *Annals of Mathematical Statistics*, 23, 1.

Zeldovich, Ya. B. 1967. "Cosmological Constant and Elementary Particles." *Journal of Experimental and Theoretical Physics, Letters*, 61, 316.

CREDITS

The author and publisher gratefully acknowledge permission to reprint the following material:

Art

Figures 4, 5, 6, 12, 13, 15, 19, 21, 25, 28: by Pam Jeffries.

Figure 18: Courtesy of the Archives, California Institute of Technology.

Figures 22, 23, 29, 30: By permission of the Master and Fellows of St. John's College, Cambridge.

Figures 32, 34, 35: Einstein, Albert; The Collected Papers of Albert Einstein. © 1987–Current Year. Hebrew University of Jerusalem and Princeton University Press. Reprinted by permission of Princeton University Press.

Figures 9, 20: Courtesy of Institute of Astronomy, University of Cambridge, through the assistance of Mark Hurn.

Figure 16: Courtesy of the author, processed by Amanda Smith, Graphics Office, Institute of Astronomy, University of Cambridge.

Figure 31: Courtesy of Amanda Smith, Graphics Office, Institute of Astronomy, University of Cambridge.

Figures 11, 17, 33: Courtesy of Pauling Collection, Oregon State University Libraries, Special Collections and Archives Research Center.

Figure 36: Courtesy of the Leo Baeck Institute, New York.

Figures 26, 37: Courtesy of the Archives Georges Lemaître, Université Catholique de Louvain, Centre de Recherche sur le Terre et le Climat G. Lemaître, Louvain-la-Neuve, Belgique.

Figure 24: Courtesy of the Reel Poster Gallery, London.

Figure 14: Reprinted by permission from Nature Publishing Group, Macmillan Publishers Ltd: *Nature*, April 25, 1953.

Figures 1, 2, 3, 7, 8, 10: Reproduced by kind permission of the Syndics of Cambridge University Library.

Figure 27: Courtesy of the Royal Astronomical Society Library, Royal Astronomical Society Correspondence 1931.

Text

Quotes from Einstein on pages 230, 237, 239, 240, 263, 266, 267: By permission of the Albert Einstein Archives, the Hebrew University of Jerusalem.

Quotes from Hoyle on pages 157, 173, 175, 186, 202, 208, 215, 216: By permission of the Master and Fellows of St. John's College, Cambridge. Through the assistance of Mr. Geoffrey Hoyle.

Quote from Gold on page 187: By permission of the Niels Bohr Library and Archives, American Institute of Physics.

Good faith efforts have been made to contact the copyright holders of the art and text in this book, but in a few cases the author has been unable to locate them. Such copyright holders should contact Simon & Schuster, 1230 Avenue of the Americas, New York, NY 10020.

INDEX

A for Andromeda (television drama), 158

Abiogenesis, 298*n*

Accelerating expansion, 224, 226–27, 252–56

Accelerating Universe, The (Livio), 23–24

Athenaeum, 281*n*

Adams, Frank Dawson, 94–95

Admiralty Signals Establishment, 182

Albert Einstein: Philosopher-Scientist (essay collection), 266

Alpher, Ralph, 167, 168, 180, 210

American Chemical Society, *Journal of*, 112, 119

American Physical Society, 175

Amish, 34, 276*n*

Andromeda galaxy, 222, 226

Anschütz, Ludwig, 287*n*

Anthropics, 256–64

Antineutrinos, 163, 167

Aquinas, Thomas, 7

Argonne National Laboratory, 283*n*

Aristotle, 7, 16, 60

Astbury, William, 105, 111, 113–14, 120, 122, 127, 128, 132, *154*

Aston, Francis, 161

Astrophysics, 10, 68, 156, 158–59, 164, 170–71, 180, 182–83, 219, 235, 294*n*

  of big bang versus steady state model, 200–202, 213

  cosmological constant in, 247

nuclear, 100–101, 159, 161–62, 171, 173–74, 178, 181, 184

principle of homogeneity and isotropy in, 185

Atkinson, Robert, 166

Atomic number, 163

Australia, 203, 207

Avery, Oswald, 117–19

Avery, Roy, 118

Baade, Walter, 169, *181*, 204–5

Bakus, Carl, 283*n*

Bamford, Clement, 112

Bardeen, John, 235

Barnes, Charles, 173

Baryonic matter, 163

Becker, Lydia Ernestine, 35, 276*n*

Beckett, Samuel, 157

Becquerele, Henri, 92, 283*n*

Béguyer de Chancourtois, Alexandre-Emile de, 293*n*

Beighton, Elwyn, 127

Bell, Florence, 120, 122, 128, 132

Bell, Jocelyn, 182

Bell Telephone Laboratories, 210

Bergerac, Cyrano de, 64

Bergmann, Peter, 237

Berkeley, George, 16

Bernal, John, 129, 151

Bethe, Hans, 100–101, 166–70

Bible, Genesis, 6, 61–62

Bierce, Ambrose, 136

Big bang theory, 157–58, 187, 210–19, 261, 262

Big bang theory (*Continued*)
nucleosynthesis in, 167–69, 219,
232, 237, 294*n*
steady state cosmology versus,
183, 199–202, 205–7, 210–19
see also Expanding universe
Biodiversity, *see* Diversity, biological
Biology, 8, 60, 82, 120, 122, 156, 271
evolutionary, *see* Evolution
molecular, 33, 114–15
Birkbeck College, 144, 151
*Biston betularia betularia morpha
typica*, 32
Black holes, 209, 263
Blending heredity, 38–44, *43*, *45*, 47,
49–50, 52, 56
Block, David, 193, 297*n*
Bluhm, Michael, 134
Bock, Fedor von, 7
Bohr, Niels, 281*n*
Bondi, Hermann, 182–83, 186–87,
198, 199, 201–3, 206–7, *209*,
211, 295*n*
Bonnet, Charles, 8
Born, Max, 203, 263
Bragg, Lawrence, 110–14, 124, 125,
141, 144, 152, *154*
Bragg, William Henry, 110, 111, 144
Branching, *see* Speciation
Branson, Herman, 109, 112, 285*n*, 286*n*
Brehm, Jack, 97
Brenner, Sydney, 124–25
British Association for the Advance-
ment of Science, 80, 81, 85, 93,
164, 280–81*n*, 283*n*
British Broadcasting Company
(BBC), 157
British Institute of Physics, 83
Brownian motion, 264
Brunauer, Stephen, 235
Brünn Natural History Society, 53
Brussels Scientific Society, Annals of,
191

Buffon, Georges-Louis Leclerc,
Comte de, 65, 278*n*
Bunsen, Robert, 293*n*
Burbidge, Geoffrey, 178–82, *179*,
181–82, 207, 212, 214, 294*n*
Burbidge, Margaret, 170, 178–82,
179, 207, 214, 294*n*
Burton, Robert, 99
Butler, N., *179*
Byron, Lord, 275*n*

Caesar, Julius, 16
Calder, L., 300*n*
California, University of, Berkeley,
119, 155
California Institute of Technology
(Caltech), 103, 106, 109, 115,
120, 131, 132, 134, 285*n*, 289*n*
Kellogg Radiation Laboratory,
166, 171, 173–74, 181
Cambridge University, 61, 134, 136,
152, 208, 279*n*, 280*n*
Cavendish Laboratory, 109–12,
125, 140–41, 161, 203, 286*n*
Darwin Correspondence Proj-
ect, 52
Franklin at, 120–21
Hoyle at, 158, 170, 178, 183
Institute of Astronomy, 179, 214,
219
Kelvin at, 67–68
King's College Biophysics
Research Unit visit to, 124
Canyon Diablo meteorite, 283*n*
Carbon-nitrogen (CN) cycle, 166
Carter, Brandon, 258, 259
Castle, William Ernest, 58
Catholic Church, 56
Chabad, 97
Chamberlin, Thomas, 75
Chambers, Robert, 20
Chandrasekhar, Subramanyan, 181
Chandrasekhar mass, 301*n*

Chargaff, Erwin, 123, 130, 138, 141–43, 145
Chase, Martha, 131
Christianity, 61, 278n
Christian Science Monitor, 205
Christie, Julie, 158
Clayton, D. D., 179
Cold Spring Harbor Laboratory, 124
Cold War, 128
Colgate, S. A., 179
Columbia University, 102
Common ancestry, 18, 21–23, 275n
Communists, 129
Condon, Edward, 166
Constitution, US, 9
Cookson, Dr., 279n
Copenhagen, University of, 120
Copernicus, Nicolaus, 24–25, 259
Corey, Robert, 104, 112–13, 154, 291n
    DNA research by Pauling and, 132–33, 142, 149, 152, 291n
    Royal Society meeting attended by, 128, 129
    X-ray crystallography studies of peptides and amino acids by, 106, 138, 285n
Cornell University, 171
Correns, Carl, 53
Coryell, Charles D., 285n
Cosmic microwave background, 168, 185, 210–11, 213–14, 254
"Cosmological Considerations in the General Theory of Relativity" (Einstein), 222
Cosmological constant, 223, 229–65, 301n
    in accelerating universe, 252–56
    anthropic reasoning on, 256–64
    Einstein's repudiation of, 237–43, 247, 267
    Gamow's account of Einstein's

"biggest blunder" remark on, 231–37
    precursors of, 246–47
    quantum mechanics and, 249–52
    scientists' continuing advocacy of, 243–45, 300n
    in static universe, 223–25, 248
    theoretical objections to, 229–31
Cosmological principle, 185, 198
Cosmology, 2, 159, 183–85, 203, 210–11, 235, 242, 257
    big bang, see Big bang theory
    Einstein on, 265–66
    inflationary, 244
    steady state, see Steady state theory
    see also Cosmological constant
Coulomb force, 165, 169, 294n
Courtaulds Research Laboratories, 112
Creation field, 200, 218
Creationism, 20, 26, 32
    see also Intelligent design
Creation myths, 9
Creation of the Universe, The (Gamow), 168
Crick, Francis, 115, 131, 154, 155
    discovery of structure of DNA by Watson and, 103, 120–28, 134–37, 139–42, 144–53, 287n, 288n, 292n
Critchfield, Charles, 166
Croll, James, 80, 81
Crommelin, Andrew, 228
Curie, Pierre, 92, 283n
Curtis, Heber, 222
Cuvier, Georges, 66
Cygnus, 204–5

Dark energy, 255, 300n
Darlington, Cyril Dean, 58
Darwin, Charles, 10, 12, 16–36, 44, 47–59, 158, 212, 230, 269–71, 277n, 281n, 298n

Darwin, Charles (*Continued*)
  and age-of-Earth controversy,
    78–82, 101–2, 202
  common descent concept of, 18,
    21–22, 275*n*
  and Copernican principle, 24–25
  gradualism concept of, 18, 20–21,
    66–67
  Huxley's support of, 81–82, 116
  inclusion of humans in evolution-
    ary theory of, 25–26, 274*n*
  Jenkin's criticism of, 39–41, 44,
    47–49, 52, 58, 59
  lack of familiarity with Mendel's
    work, 52–54
  mathematical errors of, 44, 47
  Mendel influenced by, 54–57
  on natural selection, 26–29, 31–33,
    35–36, 43, 44, 269
  and nineteenth-century theory of
    heredity, 37–38
  pangenesis theory of, 50–51
  reductionism of, 25
  speciation concept of, 18, 22–23
  works of, *see titles of specific works*
Darwin, Erasmus, 20
Darwin, George Howard, 78–79, 87,
    92, 158, 280*n*
Davis, Arthur Sladen, 40, 49
Dawkins, Richard, 217
Dead of Night (film), 198
  poster for, *186*
Death instinct, 7
De Maillet, Benôit, 63–64
Denial, 217–18
Dennett, Daniel, 26
Descartes, René, 63, 65, 263
*Descent of Man and Selection in
    Relation to Sex, The* (Darwin),
    18, 274*n*
De Sitter, Willem, 229–30, 237, 262,
    265
de Vries, Hugo, 53

Dicke, Robert, 210
Different Approach to Cosmology, A
    (Hoyle), 212
Dinosaurs, 20, 30
  feathered, 275*n*
Dirac, Paul, 159, 281*n*
Diversity, biological, 12–14
  natural selection as mechanism
    for, 18, 34
  speciation and, 22
DNA (deoxyribonucleic acid) 41, 59,
    102, 116–51, *146*, 153–55, 287–
    88*nn*, 291–92*nn*
  cosmic origins of, 170
  early work on, 117–19, 288*n*
  as evidence in criminal justice sys-
    tem, 6, 217
  Pauling's work on, 116, 119–20,
    122–23, 126–28, 130–44, 288*n*,
    291*nn*
  Watson and Crick discover struc-
    ture of, 103, 120–28, 134–37,
    139–42, 144–53, 287*n*, 288*n*,
    292*n*
  X-ray diffraction images of, 120–29,
    126, 132, 133, 138–40, 150–52
Dobzhansky, Theodosius, 20
"Don Juan" (Byron), 275*n*
Donohue, Jerry, 134, 145, 291*n*
Draper, John William, 281*n*
Drosophila fruit fly, 58
Dryster, Frederick, 281*n*
Dunbar, Noel, 173, 174
Dunitz, Jack, 112, 139, 143, 144, *154*
Dynamics, 235
  molecular, 287*n*
  see also Thermodynamics

Earth, age of, 60–83, 199, 240, 281*n*
  Kelvin's calculation of, 59, 69–79,
    81, 84–87, 91–102, 164, 210,
    280*n*
  Newton on, 64–66

Eddington, Arthur, *162*, 191, 195–96, 221, 227, 293*n*
  cosmological constant advocated by, 224, 243–44
  hydrogen fusion concept of, 100, 161–62, 164–65
Effects of Cross and Self Fertilisation in the Vegetable Kingdom, The (Darwin), 53
Eggleton, Peter, 213–15
Einstein, Albert, 7, 10, 101, 185, 187, 220–45, *242*, 262–71, 281*n*
  assumption of large-scale homogeneity and isotropy of space concept of, 185, 262
  and cosmological constant, 223–25, 229–33, 252, 255–57, 264–65, 267, 298*n*, 300*n*, 301*n*
  and discovery of expanding universe, 187, 189–91
  Eddington and, *162*
  general relativity theory of, *see* General relativity
  Lemaître and, *250*, 299*n*
  special relativity theory of, 220, 225–26, 264
Elements
  formation of, 161–66 (*see also* Nucleosynthesis)
  periodic table of, *160*, 160–61, 163, 293*n*
"Einstein's Greatest Blunder" (Leahy), 241
Elliot, John, 158
Elliott, Arthur, 112
Ellis-van Creveld syndrome, 34, 276*n*
Empedocles, 160
Empty space, 208, 247–52, 255, 257, 262
Encyclopaedia Britannica, 52, 53
England, 67, 120, 129–31, 148, 158, 161, 203
  Christian chronologists in, 61
  industrial revolution in, 32
  late-Victorian, 40
  in World War II, 182
  see also Cambridge University; Oxford University
England, Philip, 95
Eniwetok Atoll, hydrogen bomb test on, 178
Equilibrium, 167, 288*n*
  convective, 90
  statistical, 169
  unstable, 224
Escobar, Pablo, 243
Essay on the Principle of Population, An (Malthus), 275*n*
Eternal inflation, 161
Evolution, 1, 8–10, 12–36, 114, 202–3
  age of Earth and, 59, 66–67, 72, 80–81, 93
  cosmic, 10, 18, 156, 176, 180, 183–84, 206, 254, 263 (*see also* Big bang theory; Expanding universe)
  intelligent design versus, 18, 26, 30, 80
  mechanisms of, *see* Heredity; Natural selection
"Evolutionary Universe, The" (Gamow), 232
Expanding universe, 187–98, *188*, 196, 231
  acceleration of, 224, 226–27, 252–56
  discovery of, 187, 190, 192–93, 197, 223, 231
  general relativity and, 199, 244
  see also Big bang
"Expanding Universe, The" (Lemaître), 198
Expression of the Emotions in Man and Animals, The (Darwin), 274*n*

Extinction, 20, 22, 23, 30, 32, 33, 154
    population size and, 40–41
Eyring, Henry, 235

Faulkner, John, *179*, 213–15
Federal Bureau of Investigation, US,
    155
Fermi, Enrico, 129, 168–69, 294*n*
Festinger, Leon, 96
*Few and New Observations, upon
    the Book of Genesis, A* (Light-
    foot), 61
Feynman, Richard, 226, 281*n*
Field equations, 224–25, 230, 231, 247
Finlay-Freundlich, Erwin, 185
Finsbury Technical College, 281*n*
Fischer, Bobby, 5
Fisher, R. C., *179*
Fisher, Ronald, 58
FitzGerald, George, 282*n*
Focke, Wilhelm Olbers, 53
Fölsing, Albrecht, 235
Fontenelle, Bernard le Bovier de, 63
Forbes, James David, 71
Fourier, Joseph, 71–72
Fowler, P. H., *179*
Fowler, Willy, 171, 173, 178–82, *179*,
    181, 207
France, 129, 131
    in World War I, 158
Francis Joseph I, Emperor of Aus-
    tria, 56
Franklin, Rosalind, 120–21, 123–26,
    129–30, 139, 144–45, 148, 150–
    52, 287–88*nn*
Freud, Sigmund, 7, 217–18
Frič, Antonén, 56–57
Friedmann, Aleksandr, 190, 221, 230,
    238, 239, 267
Frisch, Otto, 169

Galaxies, 25, 211, 221, 251, 257,
    259–62
    active, 209, 213
    and cosmological constant, 231,
        243, 270
    in expanding universe, 187–92,
        222, 252–54
    in steady state universe, 184–85,
        198–99, 201–7
    see also Milky Way galaxy
Galen of Pergamum, 216
Galileo Galilei, 16, 62, 83, 265, 281*n*
Gamow, George, 155, 156, 165–66,
    292*n*, 294*nn*
    big bang theory conceived by,
        167–68, 180, 201, 210, 232
    Einstein and, 232–37, 241, 267,
        300*n*
    Hoyle's role in theory of forma-
        tion of elements recognized by,
        175–76
Gann, Alexander, 124
Geikie, Archibald, 75, 80
General relativity, 24, 62, 220–29,
    264, 268, 269
    black holes predicted by, 263
    and cosmological constant, 222–
        25, 239–41, 247–49, 265
    Eddington's advocacy of, 162
    in expanding universe, 189–91,
        222, 244, 252
    space-time in, 189, 220–22, 225–29
    in steady state cosmology, 199–
        200, 244–45
Genetic drift, 33–35
Genetics, 20, 30, 33
    modern, 41–42, 50
    population, 66
    see also DNA: Heredity
Geological record, 21, 66
Geological Society of Glasgow, 76
Geological Society of London, 82
Geological Survey of Scotland, 80
Geologist, 275*n*
George VI, King of England, 203

George Washington University, 155

Germany, 57, 166

  in World War II, 7, 182, 233

Gilbert, W. S., 37

Glasgow University, 67, 68, 79

Goertzel, Ted and Ben, 286n

Gold, Thomas "Tommy," 182–83, 186–87, 198, 199, 201–4, 206, 209, 295n

Gondoliers (Gilbert and Sullivan), 37

Gosling, Raymond, 124, 126, 144, 148, 150–51

GPS satellites, 229

Gradualism, 18, 20–21, 58, 66

Gravitation, 185, 221, 227, 237, 240

  and Sun's energy, 74, 100, 172, 291n

    Newton's law of, 43, 226, 246–48

    repulsive force of, 230, 246, 248

Greaves, William, 203

Greece, ancient, 229, 242

  mythology of, 6, 7

Gribbin, J. R., 179

Gribbin, M., 179

Guggenheim Foundation, 134, 141

Gurney, Ronald, 166

Guth, Alan, 218

Habitable zones, 261

Haldane, J. B. S., 26, 58

Hale, George Ellery, 227

Halley, Edmond, 206

Harding, R., 179

Heaviside, Oliver, 282n

Hegel, Georg Wilhelm, 225

Heisenberg, Werner, 281n

Helmholtz, Hermann von, 74, 100, 164

Heredity

  blending, 38–44, 43, 45, 47, 49–50, 52, 56

  Mendelian heredity, 43, 46, 47, 49–58, 115

Herman, Robert, 168, 210

Hershey, Alfred, 131

Hewish, Antony, 182

High-Z Supernova Search Team, 253

Hinduism, ancient, 60, 278n

Histoire Naturelle, Générale et Parculière (Buffon), 65, 278n

Hoffmann, Hermann, 53

Home Is Where the Wind Blows (Hoyle), 208, 215

Homogeneity, 184–85, 218, 262

"Homogeneous Universe of Constant Mass and Increasing Radius Accounting for the Radial Velocity of Extra-Galactic Nebulae" (Lemaître), 191, 195

Hooker, Joseph Dalton, 50

Houtermans, Fritz, 166

Hoyle, Barbara, 208

Hoyle, Fred, 10, 156, 172–84, 179, 181, 198–220, 209, 219, 269–71, 293nn, 295n

  background of, 158–59

  big bang theory rejected by, 157–58, 213–15

  and cosmological constant, 244–45

  idiosyncratic ideas on origin of life of, 216–17, 219, 298n

  steady state cosmology of, 183, 184, 186–87, 191, 192, 198–213, 218, 244, 247, 295n

  on stellar nucelosynthesis of, 101, 169–82

Hsien Wu, 285n

Hubble, Edwin, 197–99, 202, 209, 253

  age of universe implied by observations by, 240, 243

  discovery of cosmic expansion by, 187, 189–93, 222, 223, 231, 239, 252, 267

Hubble constant, 217, 254, 297n
Hubble Space Telescope, 189, 193, 231
Hugo, Victor, 6
Human Genome Project, 153
Humason, Milton, 189, 192
Hume, David, 16
Hutton, Frederick Wollaston, 31, 275n
Hutton, James, 21, 66, 76, 279n
Huxley, Thomas Henry, 49, 57, 81–82, 84, 116, 281n

Illustrations of the Huttonian Theory of the Earth (Playfair), 76–77
Immutability, dogma of, 16
Indiana, University of, 120
Industrial revolution, 32
Infeld, Leopold, 266
Inflationary universe, 218
Inheritance, *see* Heredity
Intelligent design, 15–16, 217, 282n
    evolution versus, 18, 26, 30, 80
Interbreeding, 22
Internal Security Act (1950), 128
International Biochemical Congress, 129
International Human Genome Sequencing Consortium, 153–54
Introduction to the Theory of Relativity (Bergmann), 237
Islam, J. N., *179*
Isotopes, 94, 164, 166, 168, 171, 283n
    in stars, 175–76, 179
Isotropy, 204
    homogeneity and, 184–85, 218, 262

James, William, 157
Jenkin, Fleeming, 38–40, 43, 44, 47–49, 52, 56, 58, 59
Jesuits, 20
Jews, 182, 251, 259
    see also Judaism

Johns Hopkins University, 253
Joly, John, 92
Jones, Harold Spencer, 205, 296n
Jowett, Benjamin, 214, 297–98n
Joyce, James, 271
Judaism, 61–62
    Hasidic, 96–97
Judson, H. F., 289n
Julius Caesar (Shakespeare), 179

Kahneman, Daniel, 6, 140, 270
Kant, Immanuel, 62–63, 74
Kelvin, Lord, 10, 21, 67–102, 212–13, 269–71, 282–83n
    calculation of age of Earth by, 59, 69–79, 81, 84–87, 91–102, 164, 210, 280n
    at Cambridge, 68
    cognitive dissonance of, 96–100
    and Darwin's theory of evolution, 79–80, 202
    ennoblement of, 68
    on ether and gravitation, 246, 247
    eulogy of, 67–69
    impact on geology of, 80–83
    Perry's challenge to assumptions of, 85–92
    and radioactivity, 92–95
    on thermodynamics, 60, 279n
Kelvin-Helmholtz timescale, 280n
Kendrew, John, 110–13, 134, 137, 141
Kepler, Johannes, 260–61, 264, 265
Kettlewell, Bernard, 276n
King, Clarence, 90, 100
King, Samuel, 34
King Lear (Shakespeare), 179
King's College, London, 120–21, 125, 130, 140, 141, 144–45, 148, 151
Kirshner, Bob, 185
Kirwan, Richard, 66, 279n
Klein, Felix, 230
Kornberg, Arthur, 150
Kragh, H., 296n
Krone, Ray, 6

Laborde, Albert, 92, 283*n*
Lahav, O., 300*n*
Laloë, S., 300*n*
Lamarck, Jean-Baptiste, 20, 66
Laplace, Pierre-Simon, 74
Larmor, Joseph, 282*n*
Laue, Max von, 110
Lawrence Berkeley National Labora-
    tory, 253
Leahy, J. P., 241
Leakey, Louis, 274*n*
Leeds, University of, 127
Leeds Grammar School, 40
Leibniz, Gottfried Wilhelm, 65
Lemaître, Georges, 193–98, 201, 224,
    297*n*
    cosmological constant advocated
        by, 243, 267, 300*n*
    Einstein and, 240, *250*
    elected to Royal Astronomical
        Society, 198
    expanding universe discovery of,
        187, 191, 192, 197–98, 222, 231
    general relativity theory applied
        to universe by, 221, 239
Le Mascrier, Jean Baptiste, 63
Leonardo da Vinci, 63
Lessing, Gotthold Ephraim, 268
Levene, Phoebus, 117
Life, origin of, 18, 21, 155, 261
    Hoyle on, 214–16, 298*n*
    see also Evolution
Life magazine, 104
Lightfoot, John, 61
Locke, John, 16, 21
Lodge, Oliver, 282*n*, 283*n*
London *Times*, 93, 129
Lorentz, Hendrik, 266
Lyell, Charles, 21, 66–67, 70, 71,
    274*n*

Mach's principle, 298*n*
MacLeod, Colin, 117–18
Macmillan's Magazine, 81–82

Macroevolution, 20
Magnetic resonance imaging, func-
    tional (fMRI), 98, 99, 290*n*
Maillet, Benoît de, *see* de Maillet,
    Benoît
Maimonides, Moses, 62
Malmquist bias, 259
Malthus, Thomas, 29, 275*n*
Manchester, University of, 241
Markham, Roy, 136
Marx, Karl, 7
Mass extinctions, 20
Massey, Harrie, 204
Mathematics, 7, 80, 104, 166, 242,
    260, 263, 270
    of age of Earth and Sun, 65, 71,
        76, 78, 82–83, 87, 92
    of blending versus Mendelian
        heredity, 40–41, 47, 57, 58
    at Cambridge, 68, 279*n*
    of cosmological constant, 224,
        257, 252, 298*n*, 301*n*
    Darwin's weakness at, 44–45
    in DNA research, 128, 155–56
    of general relativity, 189, 199, 221,
        239–40, 247–48
    of selection bias, 267
    teaching methods in, 281*n*
    of X-ray crystallography, 118
Maxwell, James Clerk, 69, 281*n*
Mayr, Ernst, 36
McCarthyism, 128–31, 142
McCarty, Maclyn, 117–18
McCrea, William, 245, 249
McGill University, 93, 98
Meaning of Relativity, The
    (Einstein), 237–38
Mechanics
    of chromosomal exchange of
        genetic material, 58
    quantum, *see* Quantum mechanics
    statistical, 169
Medical Research Council, 145
Mei long, 275n

Mendel, Gregor, 42, 47, 49–58, *43, 46,* 102, 115
Mendeleyev, Dmitry, 160, 293*n*
Meredith, R. W., 275*n*
Meselson, Matthew, 143, 150
Metaphysics, 262,
Meyer, Julius Lothar, 293*n*
Microevolution, 20
Microraptor gui, 275*n*
Milky Way galaxy, 167, 184, 201, 204, 222, 226, 260, 261
Milne, Edward Arthur, 185, 263
Milner, Peter, 98
Minnesota, University of, 97
Mirsky, Alfred, 104–5, 285*n*
Miserables, Les (Hugo), 5
Mitchell, Joni, 178
Moe, Henry Allen, 134
Molecular biology, 33, 114–15
Molecular dynamics, 287*n*
Molnar, Peter, 95
Moore, James, 82
Moreau de Maupertuis, Pierre-Louis de, 20
Morgan, Thomas Hunt, 58, 102, 116
Mount Palomar Observatory, 202
Mount Stromlo and Siding Spring Observatory, 253
Mount Wilson Observatory, 169, 193
M-theory, 261
Mullard electronics company, 208, 209
Multiverse concept, 257, 259, 261–63
Mysterium Cosmographicum (Kepler), 260
My World Line (Gamow), 232–33

Nabokov, Vladimir, 23
Nanotechnology, 161
Napoléon, Emperor of France, 7
Narlikar, Jayant, 212–14, 218
National Academy of Sciences, Proceedings of, 104, 113

National Institute of Standards and Technology, 226
Natural selection, 18–21, 26–35, 38–39, 48, 269
    hereditary processes in, 41–44, 47, 49–50, 52, 54, 162
    Jenkin's criticism of, 39–41, 48–49, 58
    Kelvin's objections to, 80, 85, 210, 282*n*
    tautological arguments against, 32–33
Natural Theology (Paley), 273*n*
Nature, 40, 49, 86, 90, 92, 146–48, 150
Nature of the Chemical Bond and the Structure of Molecules and Crystals, The (Pauling), 115
Navy, US, 235–26
    Bureau of Ordnance, 233
Nazism, 182
Nernst, Walther, 249
Neuroses, female, 7
Neutrinos, 255
"New Genesis" (Gamow), 175–76
Newlands, John, 293*n*
Newton, Isaac, 9, 16, 36, 83, 246–48, 281*n*, 300*n*
    on age of Earth, 64–65
    incompatibility of Einstein's theories with, 225–27, 229, 265, 269
    tomb of, 69
Newtonian physics, 66, 228, 243–44
New York Times, The, 11, 205, 296*n*
Nicholls, C., *179*
Niessl, Gustav von, 56
Nilsson, Lars-Göran, 291*n*
Nobel Prize, 99, 266, 286*n*
    Chemistry, 7, 110, 129, 134, 286
    Economics, 6
    Peace, 7
    Physics, 110, 111, 181, 235, 239, 251, 254, 264, 283*n*
    Physiology or Medicine, 151, 291*n*

North British Review, 39–40, 48
Nottingham, University of, 161
Nubian Aquifer, 283*n*
Nuclear astrophysics, 100–101, 159,
   161–62, 171, 173–74, 178, 181,
   184
Nuclear physics, 93, 159, 173, 178,
   302*n*
Nucleosynthesis, 162
  big bang, 167–69, 219, 232, 237,
   294*n*
  stellar, 169–82, 257

Oberhummer, Heinz, 295*n*
Oedipus complex, 7
Olby, Robert, 292*n*
Olds, James, 98
"On the Age of the Sun's Heat"
   (Kelvin), 70, 74–75
Once and Future King, The (White),
   241
"On the Dynamical Theory of Heat"
   (Kelvin), 279*n*
"On the Electrodynamics of Moving
   Bodies" (Einstein), 226
"On the Origin of Life" (Kelvin),
   280*n*
On the Origin of Species (Darwin),
   18, 20, 52, 275*n*
  amendments to, 48, 274*n*
  extended geological age world
   view in, 67, 79–80, 101–2
  first edition of, 9, 17, *19*
  German editions of, 54–55, *54, 55*
  Jenkin's criticism of, 39–40, 44,
   47, 56
  mathematical errors in, 44
  natural selection described in, 26–29
  and nineteenth-century theory of
   heredity, 37
"On the Secular Cooling of the
   Earth" (Kelvin), 69–70
Oort, Jan, 181

Öpik, Ernst, 171
Ordinary Geniuses (Segrè), 235
Oster, Gerald, 287*n*, 288*n*
Oxford English Dictionary, 8
Oxford University, 57, 107, 130, 263,
   285*n*, 286*n*, 297*n*
  New Museum, 81

Paleontology, 74
Paley, William, 273*n*
Pangenesis, 50–51, 57
Panspermia, 210
Parkinson, Stephen, 279*n*
Particle physics, 251, 255
Pasadena Conference on the Struc-
   ture of Proteins, 154
Pasteur, Louis, 103
Patterson, Clair, 283*n*
Pauli, Wolfgang, 239, 249–50, 294*n*
Pauling, Ava Helen, 107–8, 142, 291*n*
Pauling, Linus, 7, 10, 145, 212–13,
   269–71
  approach and methods of, 126–27
  DNA research by, 116, 119–23,
   127–28, 129–44, 148–49, 151–
   55, 286–88*nn*, 291*n*
  Einstein and, 233
  influence on molecular biology
   of, 114–16
  passport trials and tribulations of,
   128–29, 289*n*
  protein structure research by,
   103–9, 110–14, 127, 285*nn*
  Vitamin C preoccupation of, 215
Pauling, Peter, 119, 134–35, 139, 141,
   142, 151–52
Pecker, J.-C., 300*n*
Penzias, Arno, 210
Perfect Cosmological Principle, 198
Periodic table of elements, *160*, 160–
   61, 163, 293*n*
Perlmutter, Saul, 253, 254
Perrin, Jean-Baptiste, 161, 162

Perry, John, 85–92, 95, 97–98, 102, 281*n*, 282*n*
Perutz, Max, 109–14, 120, 125, 137, 141, 145, *154*
Petrosian, Vahe, 301*n*
Pflanzen-Mischlinge, Die (Focke), 53
Phillips, John, 80
Philo Judaeus, 62
"Phospho-tri-anhydride Formula for the Nucleic Acids" (Ronwin), 119
Phylogenetic tree, 23, 275*n*
Physical Review, 168
Physics, 7, 16, 67, 120, 241, 260, 261, 271
    of accelerating universe, 252, 254
    anthropic reasoning in, 260, 261, 263, 264
    applied, 85
    atomic, 203
    of big bang, 210
    in calculations of age of Sun and Earth, 72, 75, 79–81, 83, 91, 109
    classical, 68–69, 165
    Copernican principle in, 32–33
    and cosmological constant, 245–46, 249, 257, 364
    in DNA research, 154, 155
    evolutionary nature of theories in, 269
    Newtonian, 74
    Nobel Prize in, 110, 111, 181, 235, 239, 251, 254, 264, 283*n*
    nuclear, 93, 159, 173, 178, 302*n*
    philosophy and, 62, 222
    simplicity in, 32
    symmetry in, 32, 265
    see also Astrophysics; Mechanics
Physics World, 83, 281*n*
Pioneer 10 spacecraft, 221
Pius XII, Pope, 181, *181*
Pixley, Ralph, 173
Planck, Max, 270

Plato, 60, 160, 225
Platonic solids, 260
Playfair, John, 76–77, 82
Pocket universes, 261
Poliakoff, Martyn, 161
"Polypeptide Chain Configurations in Crystalline Proteins" (Bragg, Kendrew, and Perutz), 110
Pontifical Academy of Sciences, 180
Popper, Karl, 33, 184, 201, 262, 263
Population genetics, 66
Positron-emission tomography (PET) scans, 98
Presidential Medal of Merit, 129
Princess Bride, The (film), 271–72
Princeton University, 210, 233, 300*n*
    Institute for Advanced Study, 235
Principia (Newton), 64, 246
Principles of Geology (Lyell), 66, 70, 274*n*
"Proposed Structure of the Nucleic Acids, A" (Pauling and Corey), 132–33
Prout, William, 161
Psychoanalysis, 7

Quantum mechanics, 24, 159, 163, 165, 173, 244, 249–53, 256, 263–64, 268, 270, 294*nn*
Quarks, 255
Quasars, 209–10, 214, 247, 301*n*
Queen Mary (ship), 130

Radioactivity, 69, 92–95, 98, 100, 102, 131, 283*n*
Radio astronomy, 203–7
Ramsay, Andrew, 75
Randall, John, 121, 123, 124, 144, 287–88*nn*
Rasputin, Grigory, 243
Rauscher, Prince-Bishop, 56
Rayleigh, Lord, 283*n*
Redshifts, 222, 262

Reductionism, 24, 25, 241, 275n
Rees, Martin, 214–16, 219
Reiner, Rob, 271
"Relativistic Cosmology" (Robertson), 183
Relativity (Einstein), 239
Relativity, Einstein's theory of, see General relativity; Special relativity
Renaissance, 36
Repulsion, 231, 259
  cosmic, 221–23, 232, 239
  electrostatic, 161, 165, 173
  gravitational, 248
  strength of, see Cosmological constant
Reynolds, Osborne, 282n
Rich, Alex, 130, 139–40, 152
Richter, Frank, 95
Riess, Adam, 253, 254
RNA (ribonucleic acid), 102, 117, 292n
Robertson, Howard Percy, 183
Robinson, Robert, 129
Rockefeller Institute for Medical Research, 285n
Romanes, George, 53–54
Ronwin, Edward, 119, 127, 287n
Roosevelt, Franklin D., 233
Royal Astronomical Society (RAS), 170, 183, 192–96, 198, 202, 203, 206, 209
  Monthly Notices of, 192, 193, 195, 196
Royal College of Science, 281n
Royal Institution, 94
Royal Irish Academy, 279n
  Proceedings of, 171
Royal Society of Edinburgh, 69
Royal Society of London, 128, 129, 207
  Proceedings of, 109
Royal Swedish Academy of Sciences, 176

Rüppell's Griffon, 273n
Russell, Bertrand, 157, 264, 270
Russia, invasions of, 7
Rutherford, Ernest, 93–95, 281n
Ryle, Martin, 203–8

Sagan, Carl, 158
Salisbury, Robert Cecil, Third Marquis of, 85, 282n
Salpeter, Edwin, 171–72, 176, 294n, 301n
Schmidt, Brian, 253, 254
Schneerson, Menachem Mendel, 97
Schomaker, Verner, 119, 133, 154
Schrödinger, Erwin, 281n
Schulz, Kathryn, 298n
Scientific American, 232
Scientific method, 262
Sclater, Andrew, 52–53
Seeds, William, 124
Seeger, Raymond, 235
Seeliger, Hugo von, 301n
Segrè, Gino, 235
Selection bias, 258–59
Serengeti plains, 28
Shakeshaft, John, 206
Shakespeare, William, 37, 140, 178–79
Shaw, George Bernard, 84
Shipley, Ruth B., 128–29
Silberstein, Ludwik, 162
Simplicity, 10, 24, 267
  mathematical, 239–41
  of X-ray crystallography, 110
Sky & Telescope magazine, 158
Slipher, Vesto, 191, 192, 222
Smart, William Marshall, 193–98, 194–95
Société Scientifique de Bruxelles, 193
Soddy, Frederick, 93–94, 283n
Solomon, P., 179
Solvay Conference, 152
Soviet Union, 190
Space Telescope Science Institute, 253

Space-time, 189, 221–23, 248–49, 206*n*
  stretching of, in expanding universe, 252, 261, 262
  warped, 222, 223, 225–29
Spassky, Boris, 5
Special relativity, 220, 225–26, 264
Speciation, 18, 22–23
Species
  dogma of immutability of, 16
  extinct, 20, 22, 31
  interbreeding between, 22–23
Spencer, Herbert, 29
Spinoza, Baruch de, 62
Staphylococcus aureus, 31–32
  Methicillin-resistant (MRSA), 32
State Department, US, Passport Division, 128–29
Statistical mechanics, 169
Steady state theory, 183, 184, 198–219, *200*, 247, 295*n*
  age of Earth in, 67, 70, 83
  big bang versus, 183, 199–202, 205–7, 210–19
  continuous creation of matter in, 199–200, 244–45
  cosmological principle and, 198–99
  evolution in, 201–3
  Hoyle's refusal to abandon, 211–19
  inspiration for, 186–87, 198
  radio astronomy and challenges to, 203–10
  uniformitarian assumption of, 76
Stokes, Alexander, 150
String theory, 261, 268
"Structure of Proteins, The" (Pauling, Corey, and Branson), 112
Strutt, Robert John, 283*n*
Subatomic world, theory for, *see* Quantum mechanics
Sudoku puzzles, 99

Sun, age of, Kelvin's calculation of, 70, 74, 81, 87, 89, 99–101, 109, 164
Sunday Telegraph, 212
Supernova Cosmology Project, 253
Supernovae, 16, 176, 177, 211, 253–54, 309*n*
Supersymmetry, 263
Swamping, 39–41, 47
Symbiosis, 14
Symmetry, 24, 265
  bilateral, 51
  of DNA, 139, 145–46, 300*n*
  of general relativity, 256
  in quantum mechanics, 173
  undiscovered, for hoped-for cancellation of cosmological concept, 255–56
Szekeres, P., 301*n*
Szostak, Jack, 291*n*

Tait, Peter Guthrie, 82–83, 87–89, 282*n*
Taxonomy, 21
Taylor, A. J. P., 7
Teilhard de Chardin, Pierre, 20
Teller, Edward, 129
Telliamed (de Maillet), 63
Tempest, The (Shakespeare, 37
Theophilus of Antioch, 61, 278*n*
Theory of Relativity (Pauli), 239
Thermodynamics, 70–71, 80, 279*n*
Thomson, James, 85
Thomson, Joseph John "J. J.," 283*n*
Thomson, William, *see* Kelvin, Lord
Third Programme, The (radio show), 157
Todd, Alexander, 132, 134
Tolman, Richard C., 243
Tree of life, 22, 23, *24*
Truman, Harry S., 129
Tschermak-Seysenegg, Erich von, 53
Tufts University, 26

Tunneling, 165
Turkevich, Anthony, 294n
Tversky, Amos, 270

Ulysses (Joyce), 271
Uniformitarianism, 21, 66, 76, 78, 82,
    274n
Universe, evolution of, 10, 18, 156
    see also Big bang; Expanding uni-
    verse; Steady state universe
University College, London, 204
Untersuchungen zur Bestimmung des
    Werthes von Species und Vari-
    etät (Hoffmann), 53
Urey, Harold, 129
Ussher, James, 60, 61, 278n

Van den Bergh, Sidney, 192
Vanity Fair Album, 67–68
Variation of Animals and Plants
    Under Domestication, The
    (Darwin), 50
Vatican, 56–57
    Observatory, 180
Ventral tegmental area (VTA), 98
Victoria, Queen of England, 269

Wagoner, P., 179
Wald, Abraham, 259
Wallace, Alfred Russel, 27, 29, 47, 48,
    51, 52, 275n, 277n
Watchmaker analogy, 217
Waterston, John James, 100
Watson, James, 115, 131, 154, 155,
    284n
    discovery of structure of DNA
    by Crick and, 103, 120–28,
    134–37, 139–42, 144–53, 287n,
    288n, 292n
    and Pauling's alpha-helix model
    of protein structures, 103–4
Wave theory, 287n

Weber, Robert, 90, 282n
Wegener, Alfred, 91
Weigle, Jean, 103–4
Weinbaum, Sidney, 109
Weinberg, Steven, 251, 257, 300n
Weizsäcker, Carl Friedrich von, 100–
    101, 166
Wenzel, William, 173
Weyl, Hermann, 230
Whaling, Ward, 173, 174
Wheeler, John Archibald, 229
White, T. H., 241
Wickramasinghe, Chandra, 214
Wilberforce, Samuel, 81, 281n
Wilde, Oscar, 6
Wilkins, Maurice, 120, 121–25, 137–
    37, 139, 141, 144, 147, 148, 150,
    152, 154, 287n, 288n, 290n
Wilkinson Microwave Anisotropy
    Probe (WMAP), 254
Williams, Robley, 132
Wilson, Edmund Beecher, 117
Wilson, Herbert, 150, 154
Wilson, Robert, 210
Wilson, William E., 92
Wilson, Woodrow, 9
Witkowski, Jan, 124
Woods, M., 179
World War I, 158
World War II, 7, 128, 130, 159, 169,
    182, 203, 233, 235, 259
Wright, Sewall, 58

X-ray crystallography, 105–15, 120–
    29, 126, 132, 133, 138–40, 150–
    52, 285n

Ylem, 167, 175–76

Zeldovich, Yakov, 251
Zero-point energy, 250, 251, 255
Ziegler, Anna, 151